DISCAR
D0408824

HOW TO TAME A FOX

(AND BUILD A DOG)

HOW TO TAME A FOX

(AND BUILD A DOG)

Visionary Scientists and a Siberian Tale of Jump-Started Evolution

Lee Alan Dugatkin and Lyudmila Trut

The University of Chicago Press | Chicago and London

The University of Chicago Press, Chicago 60637
The University of Chicago Press, Ltd., London

Published 2017
Printed in the United States of America

26 25 24 23 22 21 20 19 18 17   1 2 3 4 5

ISBN-13: 978-0-226-44418-5 (cloth)
ISBN-13: 978-0-226-44421-5 (e-book)
DOI: 10.7208/chicago/9780226444215.001.0001

Library of Congress Cataloging-in-Publication Data

Names: Dugatkin, Lee Alan, 1962– author. | Trut, L. N. (Lyudmila Nikolaevna), author.
Title: How to tame a fox (and build a dog) : visionary scientists and a Siberian tale of jump-started evolution / Lee Alan Dugatkin and Lyudmila Trut.
Description: Chicago : The University of Chicago Press, 2017. | Includes bibliographical references and index.
Identifiers: LCCN 2016045441 | ISBN 9780226444185 (cloth : alk. paper) | ISBN 9780226444215 (e-book)
Subjects: LCSH: Silver fox—Genetic engineering. | Domestication. | Evolutionary genetics. | Animal genetic engineering—Russia (Federation)—Siberia. | Genetics, Experimental—Russia (Federation)—Siberia. | Belyaev, D. K. | Geneticists—Soviet Union.
Classification: LCC SF405.F8 D84 2017 | DDC 636.9776—dc23 LC record available at https://lccn.loc.gov/2016045441

♾ This paper meets the requirements of ANSI/NISO Z39.48–1992 (Permanence of Paper).

Dedicated to the memory of Dmitri Belyaev,
the visionary scientist, charismatic leader,
and kind soul behind it all

Contents

Prologue: Why Can't a Fox Be More like a Dog?

Suppose you wanted to build the perfect dog from scratch. What would be the key ingredients in the recipe? Loyalty and smarts would be musts. Cute would be as well, perhaps with gentle eyes, and a curly, bushy tail that wags in joy just in anticipation of your appearance. And you might toss in a mutt-like mottled fur that seems to scream out "I may not be beautiful but you know that I love you and I need you."

The thing is that you needn't bother building this. Lyudmila Trut (one of the authors of this book) and Dmitri Belyaev have already built it for you. The perfect dog. Except it's not a dog, it's a fox. A domesticated one. They built it quickly—mind-bogglingly fast for constructing a brand new biological creature. It took them less than sixty years, a blink of evolutionary time compared to the time it took our ancestors to domesticate wolves to dogs. They built it in the often unbearable minus 40°F cold of Siberia, where Lyudmila and, before her, Dmitri have been running one of the longest, most incredible experiments on behavior and evolution ever devised. The results are adorable tame foxes that would lick your face and melt your heart.

Many articles have been written about the fox domestication experiment, but this book is the first full telling of the story. The story of the loveable foxes, the scientists, the caretakers (often poor locals who devoted themselves to work they never fully understood, but would sacrifice everything for), the experiments, the political intrigue, the near tragedies and the tragedies, the love stories, the behind-the-scenes doings. They're all in these pages.

It all started back in the 1950s, and it continues to this day, but for just a moment travel with us to 1974.

One clear, crisp spring morning in that year, with the sun shining on the not yet melted winter snow, Lyudmila moved into a little house on the edge of an experimental fox farm in Siberia with an extraordinary little fox named Pushinka, Russian for "tiny ball of fuzz." Pushinka was a beautiful female with piercing black eyes, silver-tipped black fur, and a swatch of white running along her left cheek. She had recently celebrated her first birthday, and her tame behavior and dog-like ways of showing affection made her beloved by all at the fox farm. Lyudmila and her fellow scientist and mentor Dmitri Belyaev had decided that it was time to see whether Pushinka was so domesticated that she would be comfortable making the great leap to becoming truly domestic. Could this little fox actually live with people in a home?

Dmitri Belyaev was a visionary scientist, a geneticist working in Russia's vitally important commercial fur industry. Research in genetics was strictly prohibited at the time Belyaev began his career, and he had accepted his post in fur breeding because he could carry out studies under the cover of that work. Twenty-two years before Pushinka was born, he had launched an experiment that was unprecedented in the study of animal behavior. He began to breed tame foxes. He wanted to mimic the domestication of the wolf into the dog, with the silver fox, which is a close genetic cousin of the wolf, as a stand-in. If he could basically turn a fox into a dog-like animal, he might solve the long-standing riddle of how domestication comes about. Perhaps he would even discover important insights about human evolution; after all, we were, essentially, domesticated apes.

Fossils could provide clues about when and where the domestication of species had occurred, and a rough sense of the stages of change in the animals along the way. But they couldn't explain how domestication got started in the first place. How had fierce wild animals, intensely averse to human contact, become docile enough for our human ancestors to have started breeding them? How had our own formidable wild ancestors started on the transition to being human? An experiment in real-time, to breed the wild out of an animal, might provide the answers.

Belyaev's plan for the experiment was audacious. The domestication of a species was thought to happen gradually, over thousands of years. How could he expect any significant results, even if the experiment ran for decades? And yet, here was a fox like Pushinka, who was so much like a dog that she came when her name was called and could be let out on the farm without a leash. She followed the workers around as they did their chores, and she loved going for walks with Lyudmila along the quiet country road that ran by the farm on the outskirts of Novosibirsk, Siberia. And Pushinka was just one of the hundreds of foxes they had bred for tameness.

By moving into the house on the edge of the farm with Pushinka, Lyudmila was taking the fox experiment into unprecedented terrain. Their fifteen years of genetic selection for tameness in their foxes had clearly paid off. Now, Belyaev and she wanted to discover whether by living with Lyudmila, Pushinka would develop the special bond with her that dogs have with their human companions. Except for house pets, most domesticated animals do not form close relationships with humans, and by far the most intense affection and loyalty forms between owners and dogs. What made the difference? Had that deep human-animal bond developed over a long time? Or might this affinity for people be a change that could emerge quickly, as with so many other changes Lyudmila and Belyaev had seen in the foxes already? Would living with a human come naturally to a fox that was so domesticated?

Lyudmila had chosen Pushinka to be her companion within moments of first setting eyes on her, when she was an adorable little three-week-old, frolicking with her brothers and sisters. When Lyudmila looked into Pushinka's eyes, she felt an intense sense of connection, more than with any fox before. Pushinka also showed a remarkable enthusiasm for human contact. She would wag her tail furiously with excitement whenever Lyudmila or one of the farm workers came near her, whimpering with glee and looking up eagerly at them with an unmistakable request that they stop and pet her. No one could walk by her without doing so.

Lyudmila had decided to move Pushinka into the house after she had turned one year old, mated, and was carrying a litter of pups. That way, Lyudmila would be able to observe not only how Pushinka adjusted to living with her, but whether pups born in the company of humans might socialize differently than did other pups born on the farm. On March 28, 1974, ten days before she was due to deliver, Pushinka was taken to her new home.

The 700-square-foot house had three rooms in addition to a kitchen and bathroom. Lyudmila had moved a bed, a small couch, and a desk into one room to serve as her bedroom and office combined, and she had built a den in another room for Pushinka. The third room was used as a common area, furnished with a few chairs and a table, where Lyudmila ate her meals and where, on occasion, research assistants or other visitors could gather. Pushinka would be free to roam throughout.

When Pushinka arrived early in the morning of the first day, she began racing around the house, in and out of rooms, highly agitated. Normally, pregnant foxes so close to giving birth spend most of their time lying down in their dens, but Pushinka paced and paced from one room to the other. She'd scratch at the wood chips that lined her den floor and lie down briefly, but then she'd jump right back up again and make another circuit of the house. Though she was comfortable with Lyudmila and came over to her often for some petting, Pushinka was clearly unsettled. These strange new surroundings seemed to be causing her extreme anxiety. She wouldn't eat anything all day except for a small piece of cheese and an apple that Lyudmila had brought with her for her own snack.

That afternoon, Lyudmila was joined by her daughter, Marina, and Marina's friend, Olga. The girls wanted to be there for the big move-in day. At about 11:00 p.m., with Pushinka still pacing around the house, the three of them turned in for the night, the two girls lying down on the floor under blankets next to Lyudmila's bed. To their great surprise, and Lyudmila's relief, as they began to drift off to sleep, Pushinka silently sneaked into their room and lay down right alongside the girls. Then, she, too, finally relaxed and went to sleep.

As Lyudmila would discover over the course of many months with Pushinka, the lovable little fox would not only become perfectly comfortable living with her, she would become every bit as loyal as the most loyal of dogs.

1

A Bold Idea

One afternoon in the fall of 1952, thirty-five-year-old Dmitri Belyaev, clad in his signature dark suit and tie, boarded the overnight train from Moscow to Tallinn, the capital of Estonia on the coast of the Baltic Sea. Across the waters from Finland, but at the time, a world away, Tallinn was shrouded behind the Iron Curtain that divided Eastern and Western Europe after World War II. Belyaev was on his way to speak with a trusted colleague, Nina Sorokina, who was the chief breeder at one of the many fox farms he collaborated with in developing breeding techniques. Trained as a geneticist, he was a lead scientist at the government-run Central Research Laboratory on Fur Breeding Animals in Moscow, charged with helping breeders at the many commercial fox and mink farms run by the government to produce more beautiful and luxurious furs. Belyaev was hoping that Sorokina would agree to help him test a theory he had about how the domestication of animals had come about, one of the most beguiling open questions in animal evolution.

He carried with him several packs of cigarettes, a simple meal of hard-boiled eggs and hard salami, and a number of books and scientific papers. A voracious reader, he always traveled with a good novel or book of plays or poems, along with a number of science books and papers, on his many long train rides to the fox and mink farms scattered across the vast expanses of the Soviet Union. Even as he was intent to keep up with the rush of important new findings and theories in genetics and animal behavior pouring out of labs in Europe and the US, he always made time for his love of Russian literature. He was a particular enthusiast of works reflecting on the hardships endured by his countrymen through hundreds of years of political turmoil, works that were all too relevant to the upheavals Stalin had inflicted on the Soviet Union since his ascent to power decades earlier.

Dmitri's taste in literature ranged from the cunning folktales of the country's beloved storyteller Nikolai Leskov, in which unschooled peasants often outwit their more learned superiors, to the mystical poetry of Alexander Blok, who wrote presciently shortly before the 1917 revolution that "a great event was coming." One of his favorite works was the play *Boris Godunov*, by Russia's great nineteenth-century poet and playwright Alexander Pushkin. A cautionary tale inspired by Shakespeare's Henry plays, it recounts the tempestuous reign of the popular reformist tsar, who opened up trade with the West and instituted educational reform, but also dealt harshly with his enemies. Godunov's sudden death from a stroke in 1605 ushered in the bloody era of civil war known as the Time of Troubles. That brutal period 350 years ago was mirrored in the terror and devastation Stalin had perpetuated as Dmitri was growing up in the 1930s and '40s. Stalin's purges and his ill-conceived agricultural policies produced wave after wave of famine.

Stalin had also supported a brutal crackdown on work in genetics, and in 1952 it was still a very dangerous time to be a geneticist in Russia. Belyaev followed the new developments in the field at great risk to himself and his career. With Stalin's backing, for more than a decade Trofim Lysenko, a charlatan who posed as a scientist, had wielded enormous influence over the Soviet scientific community, and one of his primary causes was a vehement crusade against genetics research. Many of the best researchers had been deposed from their positions, either thrown into prison camps or forced to resign and accept menial positions. Some had been killed, including Dmitri's older brother, who was a leading light in the field. Before Lysenko's rise to power, Russia was a world leader in genetics. A number of the best Western geneticists—such as American Herman Muller—had even made the long journey east for the chance to work with Soviet geneticists. Now Russian genetics was in a shambles, with any kind of serious research strictly prohibited.

But Dmitri was determined not to allow Lysenko and his thugs to keep him from conducting research. His work in fox and mink breeding had given him an idea about the great outstanding mystery of domestication, and it was simply too good for him not to find a way to test it.

The methods of breeding employed by our ancestors who domesticated the sheep, goats, pigs, and cows that were so vital to the development of civilization were well understood. Dmitri employed them in his work every day at fox and mink farms. But the question of how domestication had gotten started in the first place had remained a riddle. The ancestors of domesticated animals, in their wild state, would likely have simply run away in fear or attacked if a human had approached. What happened to change this and make breeding them possible?

Belyaev thought he might have found the answer. Paleontologists had argued that the first animal to be domesticated was the dog, and by this time, evolutionary biologists were sure that dogs evolved from wolves. Dmitri had become fascinated by the question how an animal as naturally averse to human contact, and as potentially aggressive as a wolf, had evolved over tens of thousands of years into the lovable, loyal dog. His work breeding foxes had provided him an important clue, and he wanted to test the theory he was still in the early stages of developing. He thought he knew what had first set the process in motion.

Belyaev was traveling to Tallinn to ask Nina Sorokina to help him get started on a bold and unprecedented project—he wanted to mimic the evolution of the wolf into the dog. Because the fox is a close genetic cousin of the wolf, it seemed plausible to him that whatever genes were involved in the evolution of wolves into dogs were shared by the silver foxes raised on the farms all over the Soviet Union.[1] As a lead scientist at the Central Research Laboratory on Fur Breeding Animals, he was in the perfect position to conduct the experiment he had in mind. Dmitri's breeding work was of such importance to the Soviet government, because of the badly needed foreign currency the sale of furs brought into the government's coffers, that he believed as long as he explained the experiment as an effort to improve the production of furs, it could be run safely.

Even so, the fox domestication experiment he had in mind was sufficiently risky that it would have to be run far away from the prying eyes of Lysenko's goons in Moscow. That's why Dmitri had decided to ask Nina to help him get it started under the auspices of her breeding program at a fox farm in faraway Tallin. He had collaborated with her on several successful projects to produce shinier and silkier furs, and he knew she was very talented. They had developed a good relationship, and Dmitri believed he could trust her and that she would trust him.

His plan for the experiment was on a scale never before carried out in genetic research, which worked primarily with tiny viruses and bacteria, or fast-breeding flies and mice, not animals like foxes, which mate only once a year. Due to the time it would take to breed each generation of foxes, the experiment might take many years to produce results, perhaps even decades, or longer. But he felt launching it was worth both the long commitment and the risk. If it did produce results, they might well be groundbreaking.

DMITRI BELYAEV WAS NOT A MAN TO shy away from danger, and he understood how to use the considerable tools he had to negotiate the treacherous waters of Stalin's rule. When World War II broke out, he immediately joined the Soviet army and fought valiantly against the Germans on the front, rising to the rank of major by war's end, though he was only twenty-eight. Both his military service and his skill in fur breeding, producing gorgeous furs that fetched high prices, had won him the trust of his government superiors, and he had developed a reputation as both a first-rate scientist and a man who knew how to get things done. Dmitri also knew how to make good use of his considerable charm, and the mesmerizing effect he had on people, to burnish his reputation.

Belyaev was a strikingly handsome man, with a strong jaw, thick coal-black hair, and penetrating dark-brown eyes. His confidence and dignified bearing lent him a commanding presence, though he stood only five feet eight inches tall. No one who worked with him, or even just met him briefly, failed to comment about the extraordinary power of his eyes when asked to describe him. "When he was looking at you," one colleague recalled, "he was looking through you, reading your mind. Some people didn't like to go to his office, not because they had done something wrong or they were afraid of being punished. They were scared by his eyes, by his gaze." Belyaev understood this effect well and he would often intently lock people in his gaze when he spoke with them. It seemed impossible to keep anything from him or to deceive him.

His demanding standards of excellence were profoundly inspiring for some of his scientific colleagues and those who worked for him, and many of them were intensely devoted to him. He gave them confidence and pushed them to do their best work, constantly probing into new avenues of inquiry with them. A believer in lively debate, he encouraged open discussion of alternative views, and he loved volleying ideas back and forth. Some of those who worked with him weren't so enamored of his leadership, however, intimidated by his intensity and unbridled energy, while others feared his disdain for any shirking of responsibility or any sort of gossip or intrigue. He knew those he could expect first-rate work from and trust and those he could not. Nina Sorokina was one of those he could have faith in on both counts.

Disembarking from his long train journey to Tallinn, Dmitri boarded a local bus heading south, traveling roads so bumpy they barely merited the name, through many tiny villages. His destination was the little hamlet of Kohila, buried deep in the Estonian forest. Not so much a village as a corporate outpost, Kohila was typical of the dozens of these industrial-scale fur farms scattered across the region.[2] Spread out over 150 acres, the farm housed about 1500 silver foxes in dozens of rows of metal-roofed long wooden sheds, each of which contained dozens of cages. The workers and their families lived a ten-minute walk away from the farm in a bare-bones settlement of drab housing units, a small school, a few shops, and a couple of social clubs.

Nina Sorokina struck a somewhat incongruous figure against the dreary backdrop of this remote outpost. She was a beautiful, dark-haired woman, also in her mid-thirties, keenly intelligent and intense about her work, commanding a powerful position for a woman in such a vital industry. A welcoming host, she enjoyed inviting Dmitri for tea in her office whenever he visited the farm. When he arrived after his long journey, they went right away to her office to talk in private. Over tea and cakes, with an ever-present cigarette dangling from his mouth, he told her what he was proposing—to domesticate the silver fox. She would not have been unreasonable to think her friend somewhat mad. Most of the foxes at the fur farms were so aggressive that when caretakers and breeders approached them, they bared their sharp canine teeth and lunged at them, snarling viciously. When foxes bite, they bite hard, and Nina and her team of breeders wore two-inch thick protective gloves that rode halfway up their forearms when they got anywhere near these animals. But Nina was intrigued, and she asked him why he wanted to attempt this.

He told her that he had been fascinated by the unanswered questions about domestication, and that he was especially taken by the puzzle as to why domesticated animals could breed more than once a year, but their wild ancestors rarely did. If he could domesticate foxes, they might also be able to breed more often, which would be very good for business. This answer was true, but it was also good cover for her and her breeding team. If anyone should ask what they were doing, they could say that they were studying fox behavior and fox physiology, which were acceptable areas of research to Lysenko, in order to see if they could increase fur quality and the number of pups born each year. How could the authorities object to that?

He didn't want to put Nina at risk by explaining more. The full truth was that if the experiment worked, it might provide the answers to many important outstanding questions about domestication in all species. The more Belyaev had researched what was known about how animals had become domesticated, the more intrigued he had become by the mysteries about it, and those were mysteries that only an experiment of the kind he was proposing would be able to solve. How else could the answer to how domestication got started possibly be found? No written accounts of this first stage of the process were available. And though fossils of the early stages of domesticates such as dog-like wolves and early versions of domesticated horses had been found, they could reveal little about how the process got going in the first place. Even if remains could eventually be found that established what the first changes in animals' physiology had been, that would not explain how and why they emerged.

A number of other puzzles about domestication also had not been solved. One was why so few animal species out of the millions on the planet had become domesticated—only a few dozen in all, most of which were mammals, but which also included a few species of fish and birds, and a few insects, including the silk moth and the honeybee. Then there was the question why so many of the changes that had taken place in domesticated mammals were so similar. As Darwin, one of Dmitri's intellectual idols, had noted, most of them developed patches of different coloring in their fur and on their hides—spots, patches, blazes, and other markings. Many also retained physical characteristics from childhood well into their adulthood that their wild cousins outgrew, such as floppy ears, curly tails, and babyish faces—referred to as the neotonic features, those that make young animals of so many species so adorable. Why would these characteristics have been selected for by breeders? Farmers raising cows, after all, had nothing to gain from their cows having black-and-white spotted hides. Why would pig farmers have cared whether their pigs had curly tails?

Perhaps these changes in the animals' characteristics had arisen not from the artificial selection process involved in breeding by humans, but through natural selection. After all, natural selection continues to operate on species after they've been domesticated, just to a lesser degree than in the wild. Animals in the wild develop all sorts of spots and stripes and other patterns in their fur and hides, which often serve the purpose of camouflaging them. The spots and patches domestic animals develop don't play this camouflaging role, though, so why would selection favor them? There must be another answer.

Another commonality among domesticated animals concerns their mating abilities. All wild mammals breed within a particular window of time each year, and only once a year. For some, that window is as narrow as a few days and for others it's weeks or even months. Wolves, for example, breed between January and March. The window for foxes is from January to late February. This time of year corresponds to the optimal conditions for survival; the young are born when the temperature, the amount of light, and the abundance of food offer them the best odds for a successful launch into the world. With many domesticated species, by contrast, mating can occur any time during the year and for many, more than once. Why had domestication led to such a profound change in the reproductive biology of animals?

Belyaev thought the answer to all of the puzzling questions about domestication had to do with the essential defining characteristic of all domesticated animals—their tameness. He believed that the process of domestication was driven by our ancestors selecting animals according to this one key trait—that they were less aggressive and fearful toward humans than was typical for their species. This characteristic of tameness would have been the essential requirement for working with the animals in order to breed them for other desirable traits. Humans needed their cows, horses, goats, sheep, pigs, dogs, and cats to be nice and gentle toward their masters, regardless of what they were trying to get from them—milk, meat, protection, or companionship. It wouldn't do to be trampled by their food or maimed by their protectors.

Belyaev explained to Nina that in his work in fox and mink breeding he had noted that while most of the minks and foxes on the fur farms were either quite aggressive or were nervous and fearful towards people, a few were quite calm when people approached them. They weren't bred to be calm, so the quality must have been part of the natural behavioral variation in a population. This, he posited, would have been true for the ancestors of all domesticates. And over evolutionary time, as our early ancestors had begun raising them and selecting for this innate tameness, the animals became more and more docile. He thought that all of the other changes involved in domestication had been triggered by this change in the behavioral selection pressure for tameness. Rather than either avoidance of humans or aggression towards them giving them the survival advantage, now being calm around humans gave them the edge. The animals living in human contact had more reliable access to food and were better protected from predators. He wasn't sure yet how selection for tameness would have caused all the genetic changes that must have happened in the animals, but he had conceived of an experiment he hoped would eventually provide the answer.

Nina was all ears. She had also observed that some foxes, though very few, were quite calm when approached, and she was intrigued by his theory. Belyaev explained the procedure he wanted Nina and her breeding team to follow. Every year, they should choose a few of the calmest foxes at Kohila at the breeding time in late January and mate them with one another. From the pups that those select foxes produced they should again choose the calmest ones and breed them. The change from generation to generation might be subtle, he noted, even difficult to identify at first glance, but they should just use their best judgment. Perhaps, he suggested, this method would eventually lead to calmer and calmer foxes, the first step in domestication.

Dmitri suggested that Nina and her breeders assess calmness by observing closely how the foxes responded when they approached their cages or put their hands up in front of them. They might even try putting a sturdy stick slowly through the bars of the cage to see whether the foxes attacked it or held back. But he would leave it up to them to work out their methods; he was confident in Nina's judgment. Nina, in return, had faith that Dmitri's idea was worth pursuing.

Before she agreed, he wanted to discuss the risks. He knew Nina understood the danger of conducting an experiment in the genetics of domestication under Lysenko, but he nonetheless emphasized to her that she must carefully consider the issue. He told her it was probably a good idea not to mention the work to others, except her team, and he offered his suggestion that if she were asked about what they were doing, she could say that the purpose of the experiment was simply to see if they could increase fur quality and the number of pups born each year.

Without a moment's pause, Nina told him she would help him. She and her team would begin right away.

NINA'S AGREEMENT TO HELP WITH THE EXPERIMENT meant a great deal to Belyaev. This work, he hoped, could be the beginning of important research, which, if he was right about domestication, might even lead to breakthrough findings. It would also be keeping the tradition of such pathbreaking work in Soviet genetics alive, which was an urgent mission for him.

Dmitri believed that his generation of researchers must revive that tradition. This experiment, he felt sure, was the best way in which he could do his part. He and his fellow geneticists couldn't allow Lysenko and his gang to hold back serious work any longer. Before long, scientists in the West were sure to crack the genetic code, figuring out how genes were constructed and how they sent messages to the cells that determined virtually everything about how animals developed and how day-to-day life is governed. Soviet geneticists must contribute to this new scientific revolution. It was time to build anew on the pioneering work in genetics that his older brother and so many of his scientific heroes had sacrificed their careers, and sometimes their lives, for.

One of those pioneers who had given their lives for the cause of genetics was a particular inspiration to Dmitri in studying domestication. Nikolai Vavilov greatly furthered our understanding of plant domestication and was also one of the world's most important botanical explorers. He traveled to some sixty-four countries collecting seeds that were vital sources of food for the world—and for Russia. In his lifetime alone, three terrible famines in Russia killed millions of people and Vavilov had dedicated his life to finding ways to propagate crops for his country. He had started collecting seeds in 1916 and his work represented a high standard of research and perseverance that Dmitri hoped to honor. Vavilov had suffered what might have been a crushing loss right at the start of his career. Returning from England during World War I, where he had studied with some of the world's leading geneticists, armed with a treasury of plant samples he planned to use in his research, his ship struck a German mine and was sunk. All the plants were lost.

Undeterred, Vavilov launched into a new research program, searching for crop varieties that were less susceptible to disease. In time, he collected domesticated plants from all around the world, which ultimately took him to the most remote jungles, forests, and mountains looking for the birthplaces of domesticated species.[3] Reputed to sleep only four hours a night, he apparently used the extra time to write more than 350 papers and numerous books, as well as to master more than a dozen languages. He wanted to be able to talk with local farmers and villagers so that he could learn everything they knew about the plants he was studying.

Vavilov's collecting adventures are the stuff of legend and began with a journey to Iran and Afghanistan, followed by visits to Canada and the United States in 1921; Eritrea, Egypt, Cyprus, Crete, and Yemen in 1926; and China in 1929.[4] On his first trip, he was arrested at the Iran-Russia border and accused of being a spy, because he had a few German textbooks with him. In the Palmir region of central Asia, he was abandoned by his guide, ditched from his caravan, and attacked by robbers. On a trip to the border of Afghanistan, when he fell as he was stepping between two train cars, he was left dangling by his elbows as the train roared along. On a trip to Syria he contracted malaria *and* typhus, but carried on. One of his biographers wrote of his superhuman intensity, "For six weeks he did not even take off his overcoat. During the day he travelled and collected. When night came he flung himself on to the floor of some native hut. . . . Dysentery afflicted him throughout his expedition but he returned with several thousand specimens."[5] Indeed, he collected more live plant specimens than any man or woman in history, and he set up hundreds of field stations for others to continue his work. His vast collection of plant species allowed him to identify eight centers of world plant domestication; in southwestern Asia, southeastern Asia, the Mediterranean, Ethiopia, Abyssinia, the Mexican-Peruvian region, the Chiloe archipelago (near Chile), the border of Brazilian and Paraguay, and one island center, near Indonesia.

Vavilov had actually befriended the young Lysenko in the 1920s, when Lysenko received national acclaim for conducting research to help increase crop yields, a mission that was so important to Vavilov. So taken by Lysenko's claims for his research in plant breeding was Vavilov at first that he went so far as to nominate him for membership in the Ukrainian Academy of Sciences. Lysenko's claims about improving crop yields were also what, tragically, brought him to Stalin's attention. His rise to power over Soviet science is a story worthy of Dmitri's beloved Pushkin.

It all started when, in the mid-1920s, the Communist Party leadership elevated a number of uneducated men from the proletariat into positions of authority in the scientific community, as part of a program to glorify the "average man" after centuries of monarchy had perpetuated wide class divisions between the wealthy and the workers and peasants. Lysenko fit the bill perfectly, having been raised by peasant farmer parents in the Ukraine.[6] He hadn't even learned to read until he was thirteen, and he had no university degree, having studied at what amounted to a gardening school, which awarded him a correspondence degree.[7] The only training he had in crop breeding was a brief course in cultivating sugar beets.[8] In 1925, he landed a middle-level job at the Gandzha Plant Breeding Laboratory in Azerbaijan, where he worked on sowing peas. Lysenko convinced a *Pravda* reporter[9] who was writing a puff piece about the wonders of peasant scientists[10] that the yield from his pea crop was far above average, and that his technique could help feed his starving country. The glowing article the reporter wrote claimed "the barefoot professor Lysenko has followers . . . and the luminaries of agronomy visit . . . and gratefully shake his hand."[11] The article was pure fiction. But it propelled Lysenko to national attention, including that of Josef Stalin.

Lysenko claimed to have conducted a set of experiments in which grain crops, including wheat and barley, produced much higher yields during stretches of cold weather after their seeds were frozen in water before planting. This method, he said, could quickly double the yield of farmlands in the Soviet Union in just a few years. In truth, Lysenko never undertook any legitimate experiment on increased crop yield. Any "data" he claimed to have produced he simply fabricated.

With Stalin as his ally, he launched a crusade to discredit work in genetics, in part, because proof of the genetic theory of evolution would expose him as a fraud. He railed against geneticists, both in the West and in the Soviet Union, as subversives, to Stalin's great pleasure. At an agricultural conference held at the Kremlin in 1935, when Lysenko finished a fire-spitting speech in which he called geneticists "saboteurs," Stalin rose to his feet and yelled, "Bravo, Comrade Lysenko, bravo."[12]

Though initially hoodwinked by Lysenko, over time, as he looked into Lysenko's claims, Vavilov became suspicious of his results, and he asked a student to conduct research to see if he could replicate Lysenko's findings. In a series of experiments conducted from 1931 to 1935, Lysenko's claims were disproven.[13] Having revealed that Lysenko was a fraud, Vavilov became his fearless opponent. In retaliation, in 1933 Stalin's Central Committee forbade Vavilov from any more travels abroad and he was publically denounced in *Pravda*, the government's mouthpiece. Lysenko warned Vavilov and his student that "when such erroneous data were swept away . . . those who failed to understand the implications" would also be "swept away."[14] Vavilov was undeterred and kept up his fight against Lysenko, and in 1939 at a meeting of the All-Union Institute of Plant Breeding he gave a talk in which he declared, "We shall go into the pyre, we shall burn, but we shall not retreat from our convictions."[15] Shortly later, in 1940, while traveling in the Ukraine, he was picked up by four men wearing dark suits and thrown into prison in Moscow. Then, the man who had collected 250,000 domesticated plant samples, had cheated death repeatedly, and had worked to solve the puzzle of famine in his homeland was slowly starved to death over the course of three years.

Dmitri had devoured Vavilov's work. He admired both the scope of Vavilov's accomplishments and his defiant defense of genetics. He hoped that the fox domestication project would help keep Vavilov's example of innovation and fortitude alive, and he expected Vavilov would have heartily approved.

Dmitri knew that his brother Nicholai would also have been an enthusiastic proponent of the fox domestication experiment, despite his own tragic fate at the hands of Lysenko. The Belyaev family had suffered many blows in the waves of brutal crackdowns that followed the 1917 revolution, but they had stayed true throughout to their convictions.

Dmitri's father, Konstantin, had been a parish priest in the village of Protasovo, with a population of only several hundred, situated in a picturesque landscape of wide meadows and lush forests a four-hour drive south of Moscow. By all accounts, the villagers adored him. The Russian authorities did not. Soon after the 1917 revolution, the government declared the state to be atheist. It cracked down hard on religion, confiscating church property and harassing believers. Dmitri's father was imprisoned repeatedly.

By 1927, when Dmitri was ten, the harassment of the clergy had so intensified that his parents were worried for his safety. They sent him away from his hometown of Protasovo to live with Nikolai, who was eighteen years his senior and was married and living in Moscow.

Nikolai had been lucky enough to enter Moscow State University before the suppression of religion would have barred him, as the son of a priest. He majored in the new field of genetics, conducting work on butterflies.

Dmitri idolized Nikolai and, whenever Nikolai was home from college, would help him catalog butterfly specimens, while Nikolai explained how these delicate creatures might help geneticists unravel such wonders as metamorphosis. When Dmitri moved in with his brother, Nikolai was studying at the Koltsov Institute of Experimental Biology and working in the laboratory of Sergei Chetverikov, one of the country's most respected and well-known geneticists.[16] Chetverikov's lab was producing many of the country's finest scientists, and Nikolai had become a favored protégé, seen by many in the research community as one of the leaders of the next era of Russian genetics. Each Wednesday the members of the Chetverikov lab would meet for tea and discuss the most recent findings. Nikolai took Dmitri to many of these meetings. The younger brother would sit in the back, fascinated by the unbridled passion of the debates, which featured a great deal of yelling, leading Dmitri to refer to them as the "yelling meetings."

Nikolai Belyaev's reputation continued to rise, and in 1928 he was offered a job at the Mid-Asian Institute of Silk Study in Tashkent, Uzbekistan, where he moved to research silkworm genetics. This was a prime appointment, as any improvement in the production of silk might prove a boon for the Soviet industry. Dmitri had hopes of following in his brother's academic path, but he was sent next to live with his older sister Olga and her family in Moscow. Because they were struggling to make ends meet for their two children, Dmitri was enrolled in a seven-year vocational program, in which he trained to be an electrician.[17] He hoped he might still pursue a university education, but when he tried to apply for admission to Moscow State University at age seventeen, he received a rude awakening. The university was no longer admitting the sons of priests. Dmitri was forced to attend a trade college instead, enrolling at Ivanova State Agricultural Academy. At least he could study biology at the agricultural school, and many top-notch scientists visited there to give lectures on the newest advances in genetics.

In the winter of 1937, Dmitri's family received the news that Nikolai had disappeared. His research on silkworm genetics had produced important results, and he'd been appointed head of a government-funded institute in Tbilisi. During a trip to Moscow to visit family and friends in the fall of 1937, Nikolai was warned that arrests of his geneticist colleagues had begun in Tbilisi. In spite of the danger, he went back for his wife and twelve-year-old son. Only many years later did the family finally learn that soon after he returned, he and his wife were arrested. On November 10, 1937, Nikolai was executed.[18] His mother searched for Nikolai's wife for years, and finally learned that she had been sent to a prison near the city of Baysk, but she could never make contact with her or find news of what had happened to her grandson.

Nikolai's disappearance and murder fueled Dmitri's commitment to repudiating Lysenko. He knew he had to take measured steps, and while he was finishing his college degree, one of his professors had become the head of a section of the Central Research Laboratory on Fur Breeding Animals in Moscow. Upon Dmitri's graduation in 1939 the professor secured Dmitri a job there as a senior lab technician, working to breed silver foxes with beautiful fur, for sale overseas. Less than a year later, World War II had broken out. Because Dmitri had distinguished himself in service, sustaining multiple life-threatening injuries in four years of intense fighting on the front, the army was reluctant to decommission him at war's end. But his fox breeding work was deemed so important by the Minister of Foreign Trade that he was released from service to rejoin the laboratory and he was eventually appointed head of the Department of Selection and Breeding. Due to the stellar reputation he had rapidly developed for the excellence of his breeding work, Dmitri felt confident that he could begin openly speaking out against Lysenko, and he did so vigorously.

In July 1948, as part of Stalin's anti-intellectualism and anti-cosmopolitanism program, a grand plan to "transform nature" was put into place by the Soviet government and Lysenko was placed in charge of all policy regarding the biological sciences.[19] Shortly thereafter, at the August 1948 meeting of the All-Union Lenin Academy of Agricultural Sciences, Lysenko presented a talk that is widely regarded as the most disingenuous and dangerous speech in the history of Soviet science, titled "The Situation in the Science of Biology," in which he once again railed against "modern reactionary genetics,"[20] by which he meant modern Western genetics. At the end of his ranting, the audience stood and cheered wildly.[21]

Geneticists at the meeting were forced to stand up and refute their scientific knowledge and practices. Those who refused were ejected from the Communist Party and lost their jobs.[22] Reading the news of the speech, Dmitri was both distraught and furious. Belyaev's wife, Svetlana, remembers the moment her husband approached her the next day at home, having just read in the newspaper about the meeting, recounting, "Dmitri was walking toward me with tough sorrowful eyes, restlessly bending and bending the newspaper in his hands."[23] A colleague recalls running into him that day and how Dmitri had fumed that Lysenko was "a scientific bandit." Belyaev began speaking out urgently about the evils of Lysenkoism to all fellow scientists, whether friend or foe.

Though protected from being fired by the importance of his fur breeding work, Dmitri was not entirely immune to Lysenko's influence. A cartoon in a Moscow magazine lampooned him, depicting him descending from the sky in a parachute with the caption "Come down to Earth," and a group of Moscow scientists sympathetic to Lysenko organized a meeting in which they lambasted the reactionary geneticists "guided by Belyaev." Dmitri appeared at the meeting and made a defiant, impassioned speech about the importance of continuing genetic research. As a result, he was banned from teaching at the Moscow Fur Institute, and the scientific papers he submitted to journals were instantly rejected. His laboratory pay was cut in half, his staff was reassigned, and he was demoted from department head to senior scientist.

Belyaev had nonetheless managed to continue to investigate genetics through his work with minks and foxes. And some of this work gave him hope that it might just be possible that the pilot experiment Nina Sorokina was running would produce significant results in shorter time than a classic interpretation of Darwin's theory of evolution would suggest. He had an idea about why so many different changes in animals—floppy ears, curly tails, and spots, the breaking of the once-a-year mating rule—came along with the process of domestication, and why they might emerge relatively quickly. He hadn't shared this with Nina Sorokina when he visited in 1952; the idea was too provisional to share with anyone yet, especially because it cut against the grain of the prevailing wisdom about the nature of evolutionary change.

Darwin had argued that evolutionary change would usually occur in small incremental steps, and that changes of the kind associated with the dramatic modifications seen in domesticated animals would take eons to accumulate. But Belyaev had noted that with the minks brought in from the wild for the breeding program, which had begun less than thirty years earlier, striking changes in the colors of their fur had emerged in just that short time. Minks in the wild have dark brown fur. But suddenly some minks had been born with beige, silvery-blue, and white fur. And this seemed to happen over and over, much more often than any geneticist could attribute to new mutations. Belyaev thought this must mean that the wild mink possessed the genes for producing these fur colors already in their genomes, but that those genes had been what he called inactive. He proposed that the change in their environment, being brought into captivity, and the new selection pressure of being bred for fur quality must have triggered these "dormant" genes to become active.

With the foxes, he had seen that white patches that had once appeared on the feet of some foxes and then stopped showing up had suddenly reappeared in later generations, but now on the faces of some foxes. Some geneticists had suggested that genes that were inactive could be "turned on" in some way, and also that genes might for some reason start producing different effects, like the change in the location of the white patches in the foxes. Dmitri thought these kinds of changes in gene activation were behind the many changes in domestication. This suggested to him that domestication could perhaps occur much more rapidly than the standard interpretation of Darwin's theory implied.

Belyaev hoped that his fox experiment might produce such rapid change. But, then again, he could be wrong and it might produce no notable results at all. That was science. He'd come up with an idea too intriguing not to pursue, he'd set the test in motion, and now all he could do was wait for some word from Nina.

2

Fire-Breathing Dragons No More

Belyaev's idea that the silver fox was a good candidate for domestication made sense. Many at the time knew that wolves and foxes had descended from a relatively recent common ancestor, so the odds seemed good that foxes would also have some of the genes that must have been involved in wolves becoming dogs. But Dmitri was well aware that genetic closeness was no assurance that the experiment would work.

One of the most puzzling things about the history of animal domestication was that multiple efforts to domesticate the close cousins of domesticated species had failed. The zebra, for example, is such a close relative of the horse that the two can sometimes be bred. This produces a hybrid zorse, if the mating is between a male zebra and a female horse, or a hebra, if the mating is between a male horse and female zebra. But despite the close genetic link to horses, the zebra has not been successfully domesticated. Many attempts were made in Africa in the late nineteenth century. The horses brought to Africa by colonial authorities were dying off due to diseases carried by the tsetse fly, but zebras were immune to many of these diseases. They were so much like horses that it seemed perfectly logical they'd make good replacements. But those who made the attempt to breed them were in for a rude awakening.

Though zebras are herbivores, who live with the wildebeests and antelopes who graze alongside them, they're also prime targets of lions, cheetahs, and leopards, and that predatory pressure instilled in them a fierce fighting spirit. They have a mighty kick. Some brave souls did nonetheless manage to train zebras to be docile enough so that they could be ridden. The flamboyant British zoologist Lord Walter Rothschild even imported a team of them to London and once made a display of them by driving a carriage pulled by four zebras to Buckingham Palace. But they resisted actual domestication. The difference is that many animals can be trained to submit to human control, but domestication involves genetic change so that animals become naturally tame, though given individuals may be less so, as with horses that can't be broken.

Deer offer another interesting case in which close relatives have responded very differently to attempts to domesticate them. Of the dozens of deer species around the world, arguably only one—the reindeer—has been domesticated. One of the last mammals to be domesticated, perhaps twice and independently by people in Russia and by the Saami people of Scandinavia, reindeer became vital to the lives of many groups of people living in the arctic and subarctic climates.[1] That no other deer species has been domesticated is especially interesting, given that they are among the wild animals that humans have long lived in the closest proximity to, and they are not generally aggressive towards us. Deer were also one of our most important sources of food for many thousands of years; we had a strong incentive to want to raise docile herds of them. But deer are generally nervous animals, and they can be aggressive if they feel their young are in danger. If a herd is spooked, they may also stampede. As with the zebra versus the horse, deer might just not have enough genetic variation in tameness to get domestication going.

Dmitri knew full well that the fox might very well turn out to be another close relative that simply couldn't be domesticated. After all, silver foxes had been bred by humans for many decades by the time he asked Nina to help him with his experiment, and most of them were anything but tame.

The silver fox was a special breed of the red fox, which is not particularly aggressive in the wild, unless backed into a corner by predators. Though red foxes have made their way into suburban areas in Europe and the United States, where they hunt small cats, they naturally prefer to stay well away from humans, and in the wild they generally hunt smaller prey. They particularly favor rodents and small birds, although as omnivores they also eat fruits, berries, grass, and grains. They don't hunt in packs as wolves do, and apart from the period right after the birth of their pups, when parents care for the pups until they are ready to set out on their own, foxes live solitary lives. They don't mate for life, instead finding a new partner every mating season. So adept are they at staying out of sight that even the red fox, with its bright orange-red coloring, can be difficult to spot in the wild.

Foxes in captivity are a different story. Most are highly aggressive when approached by caretakers, snarling at them viciously, and some are truly ferocious. A hand coming too close to a fox in a cage risks being badly bitten, so those working at fox farms, like Nina Sorokina's Kohila farm, wore those cumbersome, but necessary, thick protective gloves.

The rewards of fox farming have been worth the risks. While foxes had long been trapped for their fur, it was only in the late 1800s that people started breeding them commercially. Two enterprising Canadians decided to start a fox farm on Prince Edward Island and see whether they could breed red foxes and produce more striking colors and textures of fur. The most popular coats they produced were of a shiny, blackish-silver, and they fetched enough in the fur market that many more fox farms were started on the island. The locals called the boom the "silver rush."

Records from the London market show that, by 1910, the price of high-quality "silver fox" pelts from Prince Edward Island had shot up from a few hundred dollars per pelt to more than $2,500, and the finest breeding pairs were selling for tens of thousands of dollars.

With such riches to be made, some Russian fur breeders decided they wanted in on the action, and they imported some of the Prince Edward Island foxes. By the 1930s the Soviet Union was exporting as many silver fox furs as any country, and the Russian breeders were building the extensive network of industrial scale farms like Kohila.

Nina Sorokina and her team, which included other breeders, as well as the common workers who kept the whole operation up and running, were well aware of the aggressive responses to expect when they approached foxes to test them as Dmitri described. He suggested that they all approach the foxes in a standard way. Keeping to a limited range of behaviors would help control for differences in their gestures that might make the foxes respond differently. If one researcher approached the foxes and put his or her face up close to the front of the cage, for instance, that might elicit a different reaction than a researcher who waved a hand in front of the cage. Approaching the foxes more slowly might elicit less response than approaching rapidly.

They should always approach the foxes slowly, Nina decided, and should also open the cage slowly and reach into it slowly with some food held in the gloved hand. When they did, some of the foxes lunged at them. Most of them backed away and snarled and sneered menacingly. But about a dozen out of the hundred or so they tested each year were slightly less agitated. They certainly weren't calm, but they weren't highly reactive and aggressive either. A few would even take the food offered from the workers' hands. These foxes that didn't bite the hands that fed them became the parents of the next generation in Dmitri and Nina's pilot work.

Within three breeding seasons, Nina and her team were seeing some intriguing results. Some of the pups of the foxes they'd selected were a little calmer than their parents, grandparents, and great-grandparents. They would still sneer and react aggressively sometimes when their keepers approached them, but at other times, they seemed almost indifferent.

Belyaev was delighted. The change in behavior was subtle, and in only a handful of foxes, but they had occurred in much less time than he had expected, the blink of time on the time scale of evolution. He was now intent on expanding the pilot program into a large-scale experiment. But doing so was outside the purview of his responsibilities at the Central Research Laboratory, so he would need the approval of his higher-ups. He could tell them he was attempting to breed foxes that had especially fine fur, and that could give birth more than once a year, as he had advised Sorokina and her team to do if they were ever questioned. But even so, a large endeavor at such a prominent institution, especially right on Lysenko's Moscow home turf, ran the risk of reprisals.

Getting started might not have to wait too much longer though. In March of 1953, Stalin had died, and the political winds were shifting. Lysenko had begun to lose his grip on power. Though Stalin's successor, Nikita Khrushchev, was also a fan of Lysenko, he was promoting a revitalization of Soviet science, which included the reinstatement of some prominent geneticists, who had been toiling as the equivalent of laboratory technicians under Lysenko's reign, back to their scientific posts. Another clear sign of changing winds was the government's official rehabilitation of the reputation of Belyaev's hero, Nikolai Vavilov.[2] There was much catch-up work to do.

Just the month before Stalin's death, James Watson and Francis Crick had announced that they had solved the vexing mystery of the structure of DNA and cracked the genetic code. Displaying a massive model of the molecule, they revealed that it was shaped like a spiraling staircase, a structure that came to be called the double helix. DNA was like a microscopic computing machine, and this discovery at last offered a compelling explanation of how mutations happen; they must arise from errors made in the copying of the code.

In light of this masterful explanation of the genetic code, Lysenko's railings against "Western genetics" were exposed as, at best, ludicrously misinformed. On top of that, many efforts to improve crop yields using methods proposed by Lysenko had failed miserably. The yields of crops from seeds produced according to his recommendations had not increased. Many experiments with grafting had also been undertaken, as Lysenko had asserted that the combination of characteristics achieved by this method would be inherited in the offspring of those hybrids. This had also proved unfounded. By stark contrast, Western scientists were producing bumper crops with their "bourgeois" genetic breeding technique of creating corn hybrids, a method Russian scientists were experimenting with in the 1930s until Lysenko had clamped down on the work.

The Soviet genetics community rallied. The leading figures in Soviet genetics, from the period of Lysenko's rise, began a bold charge in an open power struggle with the Lysenkoists. At the same time, Dmitri was gaining increasing respect in the Russian scientific community, particularly for the stunning results he was continuing to achieve in breeding beautiful animals with valuable furs. In particular, mink was becoming increasingly popular, and Belyaev had produced some glamorous new varieties of fur at the Central Research Laboratory, with gorgeous colorings of cobalt blue, sapphire, topaz, beige, and pearl. He had also written an impressive scientific paper that set forth his explanation of why some foxes had developed the white patches on their faces, due to genes that had been inactive being reactivated to produce the patches in a new location.

As word of his achievements spread, Dmitri received many invitations to lecture. His youthful energy, his eloquence, and his good looks and confidence charmed his audiences. Many who attended his lectures recall that when he walked up to a podium, he immediately commanded the whole room's attention, no matter how big the lecture hall was. Some say he possessed an almost mystical ability to sense the thoughts and mood of a crowd and to establish a strong connection with every person in the room.

On one occasion in particular, in 1954, the force of his presence and the strength of his scientific integrity made a powerful impression on the elite of the Soviet scientific community. As part of his struggle to maintain his power, Lysenko and his henchman had organized a series of lectures specifically to discredit Belyaev. They were held at the cavernous central lecture hall at the Polytechnical Museum in Moscow, one of the most prestigious venues for scientific lectures.

Dmitri was scheduled to speak, and the hall was packed for his talk. The atmosphere was electric. The crowd knew that Lysenko's cronies' purpose in inviting Belyaev to speak was to ridicule him. One of Lysenko's favored tactics was to send his minions to the public lectures of his targets to shout them off the stage with denunciations. Many lectures had famously devolved into raucous shouting matches as defenders volleyed back.

When the door to the stage opened, Dmitri strode briskly out carrying a heap of gorgeous fox and mink furs, which he draped over the lectern. As a colleague in the room that day recalled, he understood very well the effect he would achieve with this stunning visual display of his expertise. The hall fell utterly silent, and Belyaev began to speak in a deeply resonant voice. Natalia Delaunay, in the audience that day, recalled that his voice "was like a human orchestra," comparing his lecture to "a piece written for the organ."

Commanding the room, his head held proudly, his eyes locked onto those of his listeners, and they were riveted. Genetics was still an officially prohibited science, but Dmitri pulled no punches in sharing his discoveries about the genetics of breeding. He was not afraid of Lysenko and he was openly defying him. It was Belyaev who would do the ridiculing that day. And after that, he felt that he could speak openly about his disgust for what Lysenko had done to Soviet science, though he knew that those who would work with him could not.

Such respect did Belyaev command that, just a few years later, he was appointed to a high-level position that allowed him to launch the large-scale fox domestication experiment he'd dreamed of. In 1957, Nikolai Dubinin, a vocal opponent of Lysenko, was made Director of the Institute of Cytology and Genetics, one of many institutes being established as part of a gigantic scientific research center, called Akademgorodok, or "academic city." Dubinin tapped Belyaev to leave Moscow and open an evolutionary genetics lab at the Institute of Cytology and Genetics.

Part of the new push to reinvigorate Soviet science, Akademgorodok was being constructed near the large industrial city of Novosibirsk in the middle of Siberia's "Golden Valley," so called because of its abundance of natural resources. The popular concept of Siberia is of a frigid wasteland, covered in a thick blanket of snow, and it's true that the winters are brutal, with temperatures often lingering at 40 degrees below zero, but the spring and summer seasons in the Golden Valley are warm and sunny. And while vast swathes of Siberia are desolate, only dotted with tiny villages here and there, Novosibirsk was one of the largest cities in the Soviet Union, with a population close to a million, which made it a good location for a scientific hub that needed many support workers for secretarial and custodial work. As for the scientists, they would be shipped in.

Decades earlier Maxim Gorky had written of a fictional "Town of Science . . . a series of temples in which every scientist is a priest . . . where scientists every day fearlessly probe deeply into the baffling mysteries surrounding our planet." Musing of such an oasis, Gorky envisioned "foundries and workshops where people forge exact knowledge, facet the entire experience of the world, transforming it into hypotheses, into instruments for the further quest of the truth."[3]

Akademgorodok was to be such a place.

The city would house tens of thousands of researchers and would become a flourishing community of scientific comrades who would bring Soviet science to world preeminence. Not even Siberia's agonizing winters could blunt the appeal of this scientific Babylon, 2,000 miles away from Moscow and what remained of Lysenko's dwindling powerbase. Researchers, both senior and junior, from all over the Soviet Union, flocked there. They did so eagerly. It was a startling shift from the journey into obscurity, and often prison, that many persecuted scientists took during Lysenko's heyday. Now they would preside over a rebirth of science at a new scientific utopia that had been built in the most unlikely of places.

Soon after appointing Belyaev to head up the Institute's evolutionary genetics lab, Dubinin quickly promoted him to deputy director of the Institute. Dmitri would now be able to launch the full-scale fox experiment, and even before he left Moscow for Akademgorodok, he began setting the work in motion. Soon, though, he would learn that he would still need to do so cautiously.

Lysenko and his allies were furious that although they were still officially in power, geneticists on the ground were starting to simply ignore their prohibitions. They launched a new rear-guard campaign against genetics, and as part of this new battle, in January 1959, a Lysenko-created committee from Moscow arrived in Novosibirsk and visited Akademgorodok.[4] This committee had the official authority to determine what work was done at the Institute of Cytology and Genetics and who was in charge, and Belyaev and the whole research staff were at risk of being forced out. Institute scientists recall that committee members "were snooping in the laboratories," questioning everyone and anyone, including secretaries, and the word spread that the committee was clearly unhappy that genetic studies were being conducted. When the Lysenko-stacked committee met with Mikhail Lavrentyev, chief of all the institutes at Akademgorodok, they informed him that "the direction of the Institute of Cytology and Genetics is methodologically wrong." Those were ominous words from a Lysenkoist group, and everyone knew it.

Nikita Khrushchev, who was by this time Premier of the USSR, heard tell of the committee's report about its visit to Akademgorodok. Khrushchev had been a long-time supporter of Lysenko, and he decided to examine the situation personally, visiting Novosibirsk in September 1959. Khrushchev's temper often got the best of him when things did not go exactly as he ordered, and the building of Akademgorodok was a large enough project that things were not going exactly as he wanted. Indeed, he threatened to disband the whole Soviet Academy of Sciences if the situation did not improve: "I'll let you all loose!" Khrushchev railed. "I'll deprive you [of] extra pay and all privileges! Peter the Great needed an academy, what do *we* need it for?"[5]

The staff of all the science institutes at Akademgorodok gathered in front of the Institute of Hydrodynamics for Khrushchev's visit, and one researcher recalls that the Premier "walked by the assembled staff very fast, not paying any attention to them." The substance of the meeting between Khrushchev and administrators was not recorded, but accounts from the time make clear that the Institute of Cytology and Genetics would likely have been shut down by Khrushchev if his daughter, Rada, who had accompanied her father on parts of this trip, had not intervened. A well-known journalist, Rada, who was also a trained biologist, recognized Lysenko for the fraud he was, and convinced her father to keep the Institute open.

But Khrushchev decided he had to do something to show his discontent, so the day after his visit, he had Dubinin, the head of the Institute of Cytology and Genetics, sacked. As deputy director, Belyaev was promoted to take charge. Dmitri was daunted by the prospect of replacing a man as esteemed as Dubinin. But he believed in seizing opportunities even when, indeed especially when, they were challenging, and this would allow him to ensure that top-rate genetics research was conducted. A colleague and friend of his recalls that years later, when he suggested that she take charge of one of the labs at the Institute, she had told him, "I can't, I can't." She was afraid to follow in the steps of her predecessor, a woman with a stellar reputation. Belyaev told her, "Forget this expression: 'I can't.' If you want to do science, you must forget this. Do you think it was easy for me to be appointed director of this Institute after Dubinin?"[6] He took the reins, and shortly thereafter, he went in search of the person he would need to take charge of the running of his dream experiment.

"DEEP INSIDE MY SOUL," SAYS LYUDMILA TRUT, "is a pathological love for animals." She inherited this from her mother, who was a great dog lover. Lyudmila had grown up with dogs as pets, and even during WWII, when food was horribly scarce, her mother would feed starving stray dogs, telling her, "If we don't feed them, Lyudmila, how will they survive? They need people." Following her mother's example Lyudmila always carries some kind of treat in a pocket in case she encounters a stray dog. And she's never forgotten that domesticated animals *need* people. She knew that this is how we've designed them.

To follow her passion for animals, Lyudmila decided to study physiology and animal behavior, and as a top young student, she was admitted to the most prestigious program in that area in the Soviet Union, at Moscow State University, also one of the top universities in the world. Lyudmila had received the highest caliber of training of precisely the kind needed by the person who would run Belyaev's experiment. Animal behavior was an area of research with an illustrious history in Russia, and Lyudmila had learned from professors who had worked with legends.

Ivan Pavlov won the Nobel Prize in 1904 for his work on how to condition or shape behavior. Russia's first Nobelist, Pavlov showed that if dogs were always fed immediately after their keepers rang a bell, they would become conditioned to salivate at the sound of the bell, even if no food was provided. Pavlov theorized that this was a subconscious process, as opposed to a matter of conscious anticipation that food would soon be coming. His work was the foundation of the science that came to be known as behaviorism, which emphasized the effects of an animal's environment on its behavior over the role of genes in behavior. Behaviorists who followed in Pavlov's tradition included American B. F. Skinner, whose work with rats became well known in the West.

Less known was the pioneering Russian work in ethology, the study of animal behavior, which was led by naturalist Vladimir Wagner and his followers in the early twentieth century. They built upon one of Charles Darwin's core assertions, that much of animal behavior was the result of the process of natural selection. Lyudmila studied at Moscow State University, home institution to Leonid Krushinsky, one of the leading researchers who had furthered this work, and whose own work focused on the question of whether animals could think. Krushinsky was a pioneering researcher, and though he believed that genes played a powerful role in animals' behavior, he was also greatly influenced by Ivan Pavlov's work. He combined insights from both behaviorism and genetics in his research, and he advanced the view that some animals were capable of learning and elementary reasoning, and not just ruled by either genes or conditioning.

Krushinsky was inspired to study animal reasoning by observations he made of what he called animals' "extrapolation ability," by which they were able to discern where prey they were chasing had moved in order to evade them. On many trips observing animals in the wild, Krushinsky had brought his beloved dog along, and one day he observed the dog pursuing a quail into a bush. Because the bush was too thick for the dog to reach into, the dog had circled around the bush to wait for the bird to appear out of the other side. Krushinsky believed this indicated that his dog—and he was to observe this in many other animals as well—could anticipate future actions in a way that required simple reasoning. Animals must learn to extrapolate this way from experience, and surely that meant that animal behavior was shaped both by animals' genes and their life experiences and environment.[7]

As a keen investigator of the evolution of animal behavior, Krushinsky had conducted systematic comparisons of the thinking abilities of wolves versus those of dogs, claiming that the process of domestication had made dogs less intelligent. He theorized that this could have been due to the lack of survival pressure dogs were under, while wolves still needed to be constantly vigilant—keeping their wits about them, as it were—in order to survive. Since then, it has been demonstrated that dogs are actually no less intelligent than their wild cousins, and in fact have a far more diverse repertoire of behaviors than wolves or wild dogs can develop, since their lack of fear of humans allows them to adapt more readily to a complex environment.

Krushinsky also studied a host of other creatures, and extensively documented that many have complex social lives as well as problem-solving abilities. He conducted an astonishing range of fascinating studies in the field. In one paper, he wrote of his observations of the way the Great Spotted Woodpecker uses trees as tools: The birds insert pine cones into holes in trees that are just the right size to act as a kind of vise to hold the cones while they peck the seeds out of them. Although many behaviorists discounted the existence of animal emotions, and pushed the study of them to the fringe, Krushinsky wrote forthrightly about feelings he observed in animals. About African hunting dogs, for example, he noted that they live in what he called communities that are maintained by "friendly relationships."

Belyaev was friends with Krushinsky and admired his work, and because the fox experiment would require the kind of sophisticated observation of animal behavior Krushinksy taught, Dmitri visited him at his office at Moscow State's Sparrow Hill campus for advice about researchers who might be able to take charge of the daily running of the fox experiment. Ensconced in the grand setting of Krushinsky's building, with its palatial ceilings, marble floors, ornate columns, and fine art statues, Dmitri described his plans for the experiment and explained that he was looking for talented graduates to assist with the work. Krushinsky put the word out, and when Lyudmila heard about the opportunity, she was immediately captivated. Her own undergraduate work had been on the behavior of crabs, and as fascinating as their complex behavior could be, the prospect of working with foxes, so closely related to her beloved dogs, and with such a well-respected scientist as Belyaev, was tantalizing. She wanted in.

In early 1958, Lyudmila went to meet with Belyaev at his office at the Central Research Lab. She was immediately struck by how unusual he was for a male Soviet scientist, especially one of his rank. Many were quite high-handed, and condescending to women. Lyudmila, who has a genial, smiling manner and stands just five feet tall, with her wavy brown hair cropped quite short, looked young for her age, and she hadn't even finished her undergraduate studies, but Dmitri spoke to her as an equal. She was riveted, she recalls, by his piercing brown eyes that so strongly communicated his intelligence and drive, but that also emanated an extraordinary empathy. As he asked her about herself, he seemed to perceive the essence of her, as though he had known her all her life, and she felt taken into his fold.

She felt privileged to be invited into the confidence of this extraordinary man, who shared with her so openly about the bold work he was proposing. She had never experienced such a distinctive combination of confidence and warmth in a person.

Dmitri told Lyudmila what he had in mind. "He told me that he wanted to make a dog out of a fox," she recalls. Probing how creative she would be about conducting the experiment, he asked her, "You are now located on a fox farm that has several hundred foxes, and you need to select twenty for the experiment. How will you do it?" She had no experience whatsoever with foxes, and had only a vague notion of what the fox farms might be like and what sort of welcome she might receive at them. But she was a confident young women and she did the best she could to suggest some reasonable possibilities. She would try different methods, she said, talk to people who had worked with foxes, read up on what was known in the literature. He sat back and listened, gauging how committed she would be to the work and to developing techniques for such a novel study. She must be not only rigorously scientific, but also quite inventive. Was she really ready to go to Novosibirsk, to move to Akademgorodok, he asked her. After all, moving to the heart of Siberia was a life-change not to be taken lightly.

He was also clearly concerned about the risk she would be taking, and he didn't mince words about the dangers of being involved. In order to ward off the Lysenkoists, he explained, the work would be described as research in fox physiology. No mention of genetics would be made in regard to the experiment, at least for the time being. He also assured her that he could, and would, speak out against Lysenko when necessary. But Lysenko and his crowd still had the power to make an example of a team of geneticists, even those in far-off Siberia, and punish them and ruin their careers and reputations. Lyudmila knew that. Everyone knew that. Still, she was touched that he insisted that she be fully apprised.

Another serious concern he expressed was about the fate of her scientific career. He wanted to be very clear, he said, with great seriousness and looking directly and intensely into her eyes, that the experiment might not produce any meaningful results. He hoped that it would, and he believed that it would. But even if it did, that might take many, many years, even as long as the rest of her life. Her job would be to select the tamest foxes for breeding and to observe and record the details of all changes in both their physiology and their behavior from generation to generation. In addition, she would need to travel great distances away from the Institute of Cytology and Genetics in Novosibirsk to visit fox farms scattered in remote terrain, because he could not yet set up an experimental fox farm at Akademgorodok. He hoped he could one day, but not yet.

Lyudmila thought carefully about his admonitions, but she had no real doubt. This work would be a great challenge, she could see, and Belyaev would demand nothing short of excellence of her, which was greatly inspiring.

Though she was a woman of great warmth and an unassuming demeanor, Lyudmila's formidable energy and determination made her a force to be reckoned with. She had pursued her dream of becoming a scientist with great passion and had excelled at every step, despite Soviet science being almost entirely male-dominated. She wanted nothing more than to do path-breaking work. Belyaev had made it clear that she would be given a good deal of latitude and responsibility in developing her methods for working with the foxes, and that was enormously appealing. She had found, as she would later say, a "winning ticket." Not only would she be one of the first generation of researchers in a new scientific city, which might become the very center of Soviet science, but she would do extraordinary work with this remarkable man. She was sure of it. She could see it in those mesmerizing eyes of his. She trusted him.

Lyudmila had never dreamed she would leave Moscow to live in Siberia. She had grown up outside of Moscow and she loved the city. All of her family lived there, and they were very close, getting together regularly for dinners and outings. What's more, she had just married and had a baby girl. Taking her daughter, Marina, so far away from such a close circle of loving family members would be difficult. Meanwhile, who knew what sort of work her husband,

Volodya, an aviation mechanic, could find, or what sort of living conditions they could expect. The only thing she knew about living in Akademgorodok was that, being in the heart of Siberia, it would be bone-chillingly cold for much of the year. But she had to go. As it turned out, her husband heartily supported the move and felt confident he could find work there. To her great delight, her mother also decided that she would join them once they had gotten situated. She would live with them and look after the baby while Lyudmila did her work. In the spring of 1958 they took the trans-Siberian railroad and headed to their new home.

THERE WAS NO SPACE IN Akademgorodok that Belyaev could commandeer for building an experimental fox farm. The academic city was still being built, and the Institute of Cytology and Genetics didn't even have its own building yet, let alone grounds on which to house hundreds of foxes. So, at least to begin with, Lyudmila would have to conduct her work on the fox domestication experiment at a commercial fox farm. Over the years, Belyaev had developed many friendships with the managers of these farms, as with Nina Sorokina. He might have chosen to run the experiment at Kohila, but it was too small for the full-blown experiment and was also too far away. So, Lyudmila had to explore other options.

So it was that Lyudmila found herself in the fall of 1959 traveling on slow trains through vast expanses of Soviet wilderness, passing through village after village that modernity had not yet touched. She disembarked at tiny rail stations buried deep in forests and walked down dirt pathways to visit one industrial fox farm after another, looking for the best location for running the experiment.

When she arrived at a farm, she explained to the director the nature of the experiment she and Belyaev wanted to run. They'd need some space of their own and access to hundreds of foxes to test, though, she explained, they would only end up using a very small percentage of those for the breeding they'd do in their experiment, just those that were the most calm. Many at the commercial farms were mystified why anyone would want to take the time to do what Lyudmila was describing. “It is quite possible,” she recalls with amusement, “that before people knew that Belyaev had sent me, they thought I was crazy, thinking, what is she up to, wanting to pick out the tamest foxes!” But as soon as she mentioned whom she was working with, their attitude changed completely. “A single word from Dr. Belyaev,” Lyudmila recalls, “was enough to guarantee respect.”

Eventually Lyudmila settled on a giant commercial fox farm called Lesnoi, a 225-mile ride southwest of Novosibirsk, in remote terrain about halfway down to where the borders of Kazakhstan and Mongolia meet. Like all commercial farms in the Soviet Union, it was owned by the State, and at any given time, this farm housed thousands of reproductive female foxes and tens of thousands of young pups. Lesnoi was a cash cow for the government, and the tiny space the director allocated for Lyudmila to keep the foxes she would breed would hardly change that. She would import about a dozen foxes from the Kohila pilot population to Lesnoi, and a few more from other commercial farms over the next few years, but most of the first group of foxes she would mate in the experiment would come from the Lesnoi population.

The Lesnoi farm took some getting used to. It was an enormous complex, with rows and rows of open-air sheds, each shed holding hundreds of cages, with one fox per cage, often pacing restlessly around. Even that wasn’t enough space, with fox cages seemingly covering every spare inch of space. The smell, especially for Lyudmila, who was a novice, was overwhelming. And the noise, especially at meal times, could often be deafening, a cacophony of yelps and screeches. The small armies of workers who fed the foxes and cleaned their cages paid little attention at first to the intense young woman methodically going about her strange testing of the foxes. They had little time for curiosity; each was responsible for the care of about 100 foxes.

Having had no prior experience with foxes, Lyudmila was taken aback at first by how aggressive they were. Becoming acquainted with these “fire-breathing dragons,” as she called them, snarling and lunging at her when she approached their cages, she found it hard to believe that they could ever be tamed. Now she understood why Dmitri had warned her that the experiment might take a very long time.

At Lyudmila's behest, the manager of Lesnoi agreed to construct some large pens for the female foxes with wooden dens built into the front corner for them to give birth in, cushioned with wood chips to make the dens comfortable for the mothers and their pups. In the wild, a pregnant female builds a cozy den for her pups-to-be, at the base of a tree, under its roots, or under a rock cleft or on a hillside, with a narrow tunnel entrance that broadens into the main den area. Once the young are born, typically in litters of two to eight, she watches over them in the den zealously, and her male mate brings her food. It was important to Lyudmila that pregnant females were provided this comfort.

The next step, in the fall of 1960, was to bring about a dozen foxes to Lesnoi from the pilot project at Kohila. Nina Sorokina and her team had bred eight generations of foxes at Kohila by this time. For the most part, the changes they had seen in the foxes were still quite subtle. A dozen of the tamest foxes were sent to Lesnoi, and in general, they were only slightly calmer than foxes at a fur farm. But two foxes, which were both from the latest breeding season at Kohila, stood out. They were noticeably calmer. When Lyudmila saw these two, she was amazed. They would even allow her to pick them up. These astonishing creatures, already so much more dog-like than other farm foxes, gave her faith that the experiment would succeed. She named them Laska ("gentle") and Kisa ("kitty"). From then on, Lyudmila gave all of the foxes born into the experiment names, with each pup's name always starting with the first letter of its mother's name. As the years went by and colleagues and caretakers joined her in her work, they took joy in selecting these names along with her.

Lyudmila's first order of business at Lesnoi was to increase the number of foxes in the study, and to do that, she would select them from the large population there. She would have to travel from Akademgorodok four times a year, starting in October, to select the calmest foxes for mating, then in late January to oversee the mating process, again in April to observe the pups shortly after birth, and finally in June, to make more observations of them and how they were maturing. Year after year. Though Lesnoi was only 225 miles away, given the state of the Soviet train system, the trip was exhausting. She would leave Novosibirsk at 11 p.m. and reach the small city of Biysk, an hour from Lesnoi, the next morning at about 11 a.m., where she caught a bus for the last leg of the journey.

Each day, starting at 6 a.m., Lyudmila made her way methodically from cage to cage. Wearing the same sort of two-inch thick protective gloves that Nina used at Kohila, she gauged how each fox reacted to her presence as she approached the cage, as she stood by the closed cage, as she opened the cage, and as she placed a stick inside the cage. Each fox was given a score on a 1 to 4 scale for each interaction, and those with the highest aggregate score were designated the calmest. She tested dozens of foxes every day, which was both physically and mentally grueling.

The majority of foxes reacted aggressively when she approached or when she put the stick into their cages. Given the chance, Lyudmila felt sure, they would have loved to rip her hand off. A much smaller number cowered in fear at the rear of their cages, also far from being calm. The smallest number stayed calm throughout, observing her intently but not reacting. She selected from that 10% of the population to become the new parents for the next generation, joining the handful of foxes that had been brought from Kohila.

Lyudmila would take a short break for lunch in the middle of the afternoon, at the little restaurant in the village which served delicious borscht, Russian meatballs, and pancakes, then she'd head back to the farm after that for several more hours of testing, and after that, in the small room she was given at the quarters of the breeding researchers on the farm, she would record every detail of her observations that day. Finally, at about 11 p.m. she would unwind with a light dinner in the kitchen, sharing stories and jokes with the others at the house. Most of her time was spent alone with the foxes, and though she was developing a rapport with them, she often felt quite lonely.

Her visit to oversee the first mating of the foxes, in January of 1960, was quite challenging. She had written a detailed plan during her October visit for which foxes to breed with which, pairing the calmest males with the calmest females while also avoiding any inbreeding. Most of the animals complied when they were brought together for mating, but some of the females rejected their proposed partners and Lyudmila had to act quickly to find another suitable mate, which was stressful. She did not want to let Dmitri down. She was out in the unheated sheds for hours and hours in temperatures that regularly dipped to −40 or −50 degrees, and she missed her husband and daughter, Marina, terribly. Though she knew her mother was taking good care of Marina, she felt horrible that she was missing so many of the exciting moments of her daughter's early development. She couldn't even call home very often, as there was no phone at the Lesnoi farm, and long-distance calls from the private phone of the director of the farm were next to impossible to arrange. The letter service between Lesnoi and Novosibirsk was also notoriously slow and unreliable.

Thankfully, her visits to Lesnoi in April and June offered compensation. Observing the fox pups as they first opened their eyes and made their way out of their dens in April was a wonderful treat. As are the young of so many animals, fox pups are adorable. When first born, they are a little bigger than the size of a human hand and weigh only about four ounces. They are entirely helpless at first, both deaf and blind, and they don't open their eyes until eighteen or nineteen days after birth. They look like little balls of puffy fur.

By their fourth week, pups in the wild begin timidly venturing forth from their dens during the day, returning to them to sleep. They stay quite close to one another at first, rolling around on top of one another playfully and nipping at one another. Their mothers keep a close watch over them. Soon they become quite rambunctious, playing more vigorously with one another, often pouncing on each other, pulling each other's tails with their mouths and biting one another's ears. By summer, mothers are finished lactating, and the dens are abandoned. As the pups continue to mature, their play becomes more aggressive, and they establish a pecking order, with one or two becoming dominant. The father and mother bring food to the pups until autumn, when they've learned to forage and hunt and are ready to fend for themselves. At that time fox families disperse, with the pups going off on their own and the mates also separating. They look for a new mate again the following January.

To simulate the normal rearing process, Lyudmila kept the pups in the experiment in their mother's pen at all times until they were two months old, and they stayed bundled up together in the den for the first month, just as in the wild. Once they started venturing out of the den, they were allowed out into a yard by the shed to play for some time each day.

Lyudmila arrived within days of their births in April, and she wrote detailed descriptions of each of them, including their fur color, size, and weight, and made note of every little step of their growth; when they opened their eyes, when they could hear, when they first began to play. By her June sortie to Lesnoi, the two-month-old pups were unbearably cute. They seemed to savor playing with one another, rolling around in the dust. When they looked up at her with their little eyes wide open, Lyudmila couldn't help but smile. She was struck by how charming these children were and marveled anew at how much animals' behavior changes as they mature.

Lyudmila felt she was making good progress getting the experiment started, and she loved her time with the foxes, but the work was taking a heavy toll on her. The long absences from her daughter continued to weigh on her, and she sometimes wondered whether she shouldn't try to find another research project, based at the Institute.

One day after her second January trip to Lesnoi, Lyudmila was waiting at the small train station in the town of Seyatel where she caught a bus to Akademgorodok. The temperature was about forty below zero, and the station was barely heated. When it was announced that there would be no buses for quite a long time, she decided that was it, she would give Belyaev her resignation the next day and her family would move away from this land. But the next morning, after a cup of hot coffee, she realized she couldn't leave. She had fallen in love with the work.

After the second mating season in January of 1961, with the birth of the second generation of pups, her experimental population of foxes included one hundred females and thirty males. As this new generation of pups matured, some of them were so comfortable with people, like the two astonishing foxes from Kohila, Laska and Kisa, that they allowed Lyudmila and caretakers at the farm to hold them. But those were the exceptions. The rest of the pups matured to be just mildly calmer than was typical of captive silver foxes, and they often still exhibited fear or aggression. They might even bite on occasion, so that gloves still had to be worn when handling them.

Lyudmila was feeling increasingly confident, though, that the experiment was working. This was due not only to the calmer behavior of more of the foxes in the newest generation, but to the change in behavior of some of the farm workers toward those calmer foxes. A few workers at Lesnoi had been assigned to help her as caretakers of the foxes, and they had begun petting the calmest foxes when they brought them their food or came to clean their pens, spending a little extra time with them and clearly forming a bond with them. One worker in particular, named Fea, had fallen in love with the calmest foxes. She was quite poor, and could barely make ends meet by working at the farm. But Fea would bring her breakfast to the farm every day and fed most of it to her favorite foxes. She loved petting them and picking them up, even when they'd become fully grown and weighed a considerable 10–20 pounds.

This kind of affection was natural with small pups, who were so adorable and docile. But to see such a strong bond forming with mature foxes was striking to Lyudmila. As an animal lover, she felt their pull too, and she occasionally allowed herself to pet them and pick them up as she was making her measurements. But for the most part she held back. She had to remain an objective scientific observer and see that others did as well, and over the years she was obsessive about that. Yet this bond that the occasional worker like Fea was forming with the foxes was an important part of the study, she felt sure. Belyaev had conjectured that our ancient ancestors selecting animals for their tameness was one of the earliest stages in getting the process of domestication rolling, and here Fea was, in real time, doing just that. It required no stretch of the imagination to envision that naturally tamer wolves who ventured into contact with our early ancestors would have elicited a similar response.

After Lyudmila returned to the Institute of Cytology and Genetics from her second June visit to Lesnoi, Belyaev and she began to analyze all of the results, pouring through the voluminous data she had collected. They were stunned to discover a change underway in some of the foxes. By visual inspection of the female foxes' reproductive organs, as well as by analyzing vaginal smears, Lyudmila had made meticulous note of when each of the females had gone into estrous each season, opening the short window of a few days in which she could mate them. Her data indicated that some of the tamer foxes were mating a few days earlier in the winter than is normal for silver foxes. Not only that, but their fertility was a little bit higher—they were producing, on average, slightly larger litters. A link between selection for tameness and more frequent reproduction was one of the pillars of Dmitri's theory that somehow selecting for innate tameness kicked off all of the changes involved in domestication. Even this slight modification of a mating cycle that had been so fixed in the species for so long seemed to be a strong indication that he was right about that link, and also that a true process of domestication, not just of breeding marginally tamer foxes, was already underway.

3

Ember's Tail

One morning in April of 1963, shortly after the fourth generation of pups was born at Lesnoi, Lyudmila was making her rounds observing them. The pups had only recently opened their eyes and left their dens. They were especially precious in these early days of exploring their world. By the time pups were three weeks old they were little bundles of energy. When they weren't being groomed by their mother, or feeding blissfully at her belly, all snuggled up next to one another in a neat little row, they were scampering around their pens, pouncing on one another, yelping gleefully and tugging each other's tails. Little fox pups are every bit as cute as dog puppies and kittens. Something about the neotonic features—the disproportionately large heads and eyes of all of these little creatures, along with their fuzzy fur and rounded little snouts—makes them irresistibly cute to humans, calling out for us to pick them up and cuddle them. On rare occasions Lyudmila gave in to that impulse and picked a little pup up. But she did her best to resist and simply observe the pups.

She visited all three dozen or so pups born to the calmest mothers several times a day, closely observing their reactions to her, how timid or bold they were, whether they were frightened if she reached in to touch them, or stayed calm, and taking detailed notes about each one on their length, size, coat color, anatomical features, and general health. As she walked up to the pen of one litter that day, one little male pup, named Ember, began vigorously wagging his tiny tail. Lyudmila felt overcome with joy. He looked just like a little dog puppy wagging his tail at her. It's really true, she thought, the foxes are becoming more like dogs! Ember was the only pup in his litter with his tail wagging, and she felt as though he were calling out to her, bursting with excitement to see her.

Wagging their tails in response to humans is one of the signature behaviors of dogs, and until that day, they were the only animals observed to do so. None of the other pups she tested had ever done this. The behavior was unheard of in foxes, either in captivity or in the wild. Foxes do wag their tails towards each other, or to rid themselves of fleas or other pests, but fox pups had not been observed to do so in response to a person approaching.

Lyudmila quickly checked her emotions. She mustn't make too much of this, she told herself, not yet. It seemed clear that Ember had started wagging his tail in response to her, but she'd have to verify that to know for sure, carefully observing whether or not he began wagging his tail again the next time she came to check on him and his siblings. Still, this was exciting. Tail wagging might be the first sign of the emergence of distinctively dog-like behavior in the foxes, and she hoped that some other pups would also wag their tails at her as she continued her rounds that morning. Not one other pup did so. Not that day, or any other day as she continued to observe them in the next couple of weeks. But Ember did keep wagging his tail, and there was no question that he started doing so when she came close to him. He also wagged his tail in response to attention from the caretakers.

Was Ember just an anomaly? Or might Belyaev and she have already found important evidence about the genetic origins of behavior in animals? Ivan Pavlov and many who had followed in the behaviorist vein of research argued that dog behaviors toward humans, including tail wagging, were the product of conditioning, which Pavlov had demonstrated with the dogs he'd conditioned to salivate at the sound of a bell. But for a new behavior to be picked up that way, an animal had to be subjected many times to a stimulus associated with the behavior. The American psychologist B.F. Skinner, one of Pavlov's most influential followers, had demonstrated a different kind of conditioning, which he called operant conditioning. This involves rewarding an animal whenever it performs a certain behavior, such as in famous experiments Skinner conducted in which he rewarded rats with a pellet of food whenever they pressed a lever with their feet. At first the rats would press the lever only by chance, but after the pellets of food appeared a number of times, they began intentionally pressing the lever. This method is used to train all sorts of animals, from dogs to seals, dolphins, and elephants. But neither type of conditioning was involved with Ember wagging his tail toward her. He had simply spontaneously started doing so. This little pup might be leading the way in the foxes displaying a newly innate dog-like trait, just as Belyaev had predicted would begin to happen. But a single animal performing a new behavior, even over and over, might just be a quirk. It would be fascinating to see if any of Ember's pups in the next generation, or any of the other pups next spring, would be tail waggers.

Lyudmila observed no other striking new behaviors in Ember's generation, but she did note that many more of the pups were markedly calmer when tested than in the prior generations. And more tame females also went into estrous a few days earlier than the normal timing for wild females, which was another good sign that the experiment was continuing to produce notable results.

She would have loved to share this news with Dmitri right away, but she'd have to wait until she returned to the Institute of Cytology and Genetics. She always had a meeting with him shortly after she got back from Lesnoi, and these meetings were special to her because they offered the rare opportunity for the two of them to discuss their findings in depth and share notions about what the results were telling them. Belyaev wished he could spend more time with Lyudmila on the fox experiment, and that he could visit the foxes regularly. But he was so busy with his work running the Institute that he had been able to sneak in only a couple of quick trips to Lesnoi so far. These meetings when she came back to the Institute with the latest news were also special to him.

He would invite Lyudmila to his office and order some of his favorite brew of tea—a special blend of Indian and Ceylon with 1.5 lumps of sugar, "every time, without exception," recalls his secretary. He'd first ask Lyudmila how her husband, daughter, and mother were faring, sensitive that her time away at Lesnoi had been hard on her family. He'd then ask how *she* was doing. Though Belyaev was an intensely driven man who worked at a feverish pace, he took the time to check in this way with those who worked for him, and he understood how difficult the long trips away had been for Lyudmila, and that she especially was missing time with young Marina, who was now a lively toddler. Lyudmila recalls how "At times when there was something wrong in my soul he [Dmitri] would feel it. And if I started to talk, well, I wouldn't even complete a word before he understood what I wanted to say."

For their meeting this time, she was delighted that she had particularly intriguing news for him. She filled him in on how calm some of the foxes were compared with earlier generations and that more females were displaying a slightly longer reproductive period. Then Lyudmila told him about Ember and his tail wagging. Dmitri agreed it could be important. Ember appeared to be wagging his tail due to a new emotional response to people, and if other pups also began to do so, that might prove to be a big step in the process of domestication. Though they'd have to wait to discover whether that was the case, the results they'd already recorded in sum were substantial enough that Belyaev decided it was time to announce them to the world genetics community. He would have the perfect opportunity for doing so, having secured a slot for a presentation at the 1963 International Congress of Genetics, which was being held at The Hague, in the Netherlands. For the first time since Lysenko had wormed his way into power decades earlier, the government was allowing a delegation of Soviet geneticists to attend this meeting, a clear sign that Lysenko was losing the power struggle. Held only once every five years, the Congress was the most important conference in genetics in the world; *the* one "don't miss" genetics meeting. Dmitri made sure he was on the list.

Over the past several years, the Russian genetics community had continued to wage its fight against Lysenko, and the wider scientific community had also taken up the cause. In 1962, three of the most respected physicists in the Soviet Union had joined in a public excoriation of Lysenko's work. He remained the director of the Institute of Genetics for another two years, but after physicist Andrei Sakharov lambasted Lysenko in 1964 in a speech to the General Assembly of the Academy of Science, blaming him for the "shameful backwardness of Soviet biology . . . for the defamation, arrest, and even death of many genuine scientists," Lysenko was deposed from his position. Shortly thereafter, the government officially denounced him and repudiated his work. Belyaev, his wife recalls, was thrilled. Soviet genetics could at last begin making up for lost time.

For his presentation at the Congress of Genetics at the Hague, Dmitri introduced the hypothesis guiding the fox experiment, about selection for tameness leading to domestication, and explained exactly how the experiment was being conducted, walking his listeners through the findings of the pilot study and then all of their latest results. The crowd was impressed; no one had heard of any domestication experiment of this kind. It was audacious. One of those who attended the talk was Michael Lerner, from the University of California at Berkeley, who was widely regarded as one of the world's leading geneticists. He introduced himself to Belyaev afterwards and the two discussed the experiment further. Lerner was struck by the scope and originality of the work, and he and Belyaev began a correspondence to keep up with one another's research. One of Dmitri's main aims in attending the Congress was to spread the word about the experiment to geneticists in the West, and Lerner couldn't have been a better man for the job. A few years later, Lerner wrote about the experiment's results in his textbook about animal breeding, one of the major works on the subject. Dmitri wrote to his friend, "I was very pleased to find references to my work."[1]

Garnering such recognition for their research outside of the Soviet block was still nearly impossible for Soviet scientists. Though they could now openly keep up with the research being done in the West, and a select number were allowed to attend some conferences abroad, the Cold War was raging and the Soviet government made submitting their work to research journals outside the Soviet block very difficult. They could sometimes sneak papers out with visitors from the West, but for the most part their work was not known there.

Belyaev was acutely sensitive to the frustration his research staff felt about this isolation. Great advances in genetics had been made in the West in recent years. Dmitri couldn't do much to help his people get their work published in the West, but at least he could facilitate their doing cutting-edge work. He worked hard to build the Institute of Cytology and Genetics into a first-rate research center, and as Dubinin had anticipated when he selected Belyaev as his right-hand man, he proved to be a strong leader who knew how to recruit top talent. The fox experiment was only one of many important ongoing projects at the Institute. Other researchers were working on basic genetics studies, such as a major project to compile an archive of the chromosomes of a host of species. Some were studying how cells function and are built. Still another group was working on crop breeding.

Dmitri was also intent to foster a spirit of camaraderie among both the staff and students at the Institute. That was made difficult by the fact that the construction of a building meant to house the Institute had been stalled for years, and the 342 staff, scientists, and students of the Institute of Cytology and Genetics had been scattered all over the grounds in five different buildings.[2] In 1964, he was at last able to bring them all together, making good use of his shrewd sense of how to negotiate political waters. When construction of a new building finally began moving forward, the increasingly powerful Computing Center at Akademgorodok lobbied hard that they deserved a nice new home more than the Institute of Cytology and Genetics, but Belyaev beat them to the punch. As soon as the building was finished, even before the ribbon cutting, he told the staff to start setting up shop in it. They moved in so quickly—in a single weekend—that before the Computing Center heads got the word, it was a fait accompli.[3]

Dmitri savored the evening, when he'd finished with his loads of administrative work and could finally turn to science. He would often invite a group of researchers or students to join him in discussions of their research. Exclaiming to his secretary, "Ok, tonight is the night, now I can do some science!," he'd have her call people to his office for a working session. This required them to work long hours, but he made it well worth their while, conducting discussions that were lively affairs. They could get quite animated, and his secretary recalls lots of shouting but also lots of laughter erupting from his office. This was exactly the way he felt scientific discussion should be conducted, reminiscent of the "yelling meetings" he attended with his brother Nicholai and the Chetverikov science group as a child.

Many of these sessions also took place at the Belyaev home, which was a short walk from the Institute. His wife Svetlana would cook a scrumptious dinner and they'd eat at about 9:00 p.m., over a heated discussion about current events. Dmitri, who was now out of his standard dark suit and tie and dressed causally, would sometimes hold forth with a story. "He was an excellent storyteller and an actor," his student, and later colleague, Pavel Borodin, recalls. "He would never just tell a story; he would play the role of the hero," doing a lively imitation. After dinner, they'd head upstairs with Dmitri to his study to talk more science and work on journal papers.

Lyudmila immensely enjoyed these sessions, and the intense debates with her colleagues about the significance of the fox experiments' intriguing findings. They were fascinated by the early results, and they shot ideas back and forth about what might be causing these changes so quickly. Soon, she would have an astonishing new set of findings to share with them.

IN 1964, LYDUMILA OBSERVED NO BIG NEW changes in the new [fifth] generation of pups. She had mated Ember to a tame female that January, in hopes that some of his pups would also wag their tails, but none did. No pups born to other females that year wagged their tails either. An increasing number of pups were markedly tamer, however.

The next generation of pups was a very different story. On her April trip to Lesnoi in 1965, to observe the newborns of the sixth generation, she discovered that they displayed a set of exciting new dog-like behaviors. These pups pressed themselves up against the front of their pens when Lyudmila approached, trying to nuzzle up against her, and rolled over on their backs, clearly inviting her to rub their bellies. They also licked her hand when she reached in to test them. When she walked away from them, these pups would whine with a sound of distress: they seemed to want her to stay. They also behaved in all of these ways with the caretakers. Just as with Ember's tail wagging, no one had ever observed these behaviors towards humans in foxes before, whether in nature or in captivity. Pups whine for food and attention from their mothers, but never had they been known to whine to solicit human attention. Nor had a fox ever been recorded to have licked the hands of caretakers. So moving were these pups' protestations that Lyudmila had a hard time disappointing them, and she often now found herself stepping back to a cage to spend a little more time before leaving. There seemed to be no doubt at all that these pups, from as early as they could walk, eagerly sought contact with humans.[4]

Dmitri and Lyudmila decided that they should designate the small number of foxes that were displaying these new behaviors as the "elite." They devised a strict categorization scheme. Class III were foxes that fled from experimenters, or were aggressive towards humans; Class II foxes were those that allowed themselves be handled, but showed no emotional response to the experimenters.

Class I foxes were those that were friendly, displaying whining and tail wagging. And the elite, Class IE, displayed in addition to those two behaviors, a distinct whimpering for attention; they sniffed and licked Lyudmila when she came to observe them, displaying clear eagerness for human contact.

Ember sired another litter the following year, and Lyudmila had hoped that the pups in this clutch would be tail-waggers, but again, none of them was, but in the next year, 1966, Ember sired a third clutch, and several of these pups did wag their tails. Ember was not an anomaly, he was a pioneer. Now Belyaev and Lyudmila had some evidence that the tail wagging was heritable.

In the seventh generation of pups, several more displayed the whining, licking, and rolling over for belly rubs behavior, but none of these pups except those from Ember's line were tail waggers. Changes were showing up differently in different litters. Something was going on in the genetic makeup of some of the tamer foxes that was leading them to spontaneously perform a whole set of brand new behaviors. And the changes were appearing in an increasing number of pups. In the sixth generation, 1.8% of the pups were elite. By this seventh generation, approximately 10% were. By the eight generation, tails were not only being wagged, but some of the tails of the tame foxes were curly, another remarkably dog-like characteristic.

That so many and such various changes in the behavior of animals could arise so early in their development was particularly notable. Natural selection stabilizes the developmental regime, and once a trait has entered this early development routine, it rarely changes, presumably because these stages of growth are so crucial in the fight for survival. That's why all fox pups opened their eyes and emerged from their dens according to a relatively fixed timeline. But the tamest pups were breaking even that rule. Lyudmila's meticulous observations revealed that tame pups were responding to sounds two days earlier and opening their eyes a day earlier than was normal. It was almost, she thought to herself, as if these little foxes were itching to start interacting with people.

As Lyudmila continued to observe the tamer pups with their new behaviors, she found that not only did they retain those new behaviors, they also held onto the characteristic puppy behaviors seen in all foxes much longer. While fox pups, like almost all animal pups, are curious, playful, and relatively carefree when they are very young, the behavior of foxes both in the wild and in captivity dramatically changes when they turn about forty-five days old. At that point, which is when wild pups begin exploring on their own more often, they become much more cautious and anxious. Lyudmila was finding that the tame pups were retaining the typical impishness and curiosity almost twice as long, for about three months, and after that, they stayed markedly calmer and more playful than is typical for foxes. These tamer foxes seemed to be resisting the mandate to grow up.

In less than a decade, the experiment had accomplished so much more than Dmitri had expected. Now it was time, he decided, to build an experimental fox farm in Akademgorodok and scale the experiment up further. Their own farm dedicated to the experiment would allow for a larger population of foxes, and Lyudmila would be able to observe them continually, not just four times a year. Belyaev could assign some research assistants and students at the Institute to help her with the work, and the Institute of Cytology and Genetics could conduct more extensive analyses of the changes underway in the foxes. What's more, Dmitri would finally be able to visit the foxes regularly himself. Due to his heavy administrative load at the Institute and many trips he had to make for conferences and lectures, he'd still only been able to make time for a few short trips to Lesnoi to see the foxes for himself. With such strong results from the Lesnoi foxes, Belyaev could now justify the allocation of the considerable funds required for building and maintaining an experimental farm. He also now had the administrative clout to do so. He started to search for property to house the farm.

ONE DAY IN MAY OF 1967, after Dmitri had poured through her data from their seventh generation of foxes, he excitedly called

Lyudmila into his office. He told her he hadn't slept at all the night before because his mind had been racing. He had an idea about what was causing the changes in the foxes, and he asked her to gather a number of the researchers at the Institute to come to his office. Once they had settled in, Belyaev told them, "My friends, I think I have come close to understanding what we are observing in the domestication experiment."

Belyaev had realized that most of the changes they'd seen in the foxes involved changes in the timing of when traits turn on and off. Many of the changes they were observing in the tamer foxes involved retaining a juvenile trait longer than normal. The whimpering was a youthful behavior that normally stopped as foxes matured. So was calmness; fox pups are serenely calm when they're first born, but as they age, foxes typically become quite high-strung. A change in timing was also going on with some of the females' reproduction systems. Their readiness for mating was occurring much earlier and was lasting considerably longer.

Hormones were known to be involved in regulating the timing of development and of the reproductive system. They were also known to regulate the levels of an animal's stress, or calmness. Dmitri felt sure that changes in the production of hormones were unfolding in the tame foxes and that this must be central to the process of domestication. If this were true, it could explain why domesticated animals look more juvenile than their wild cousins, as well as why they can reproduce outside of the normal mating time, and why they are so calm around us.

The discovery of hormones back at the dawn of the twentieth century had shaken the foundation of animal biology. The basic operation of the nervous system was just starting to be pieced together at that time, and the brain and the nervous system were thought to be the communication system that regulated animal behavior. Then, suddenly, it seemed our bodies were also controlled by a chemical messaging system, and it operated through the bloodstream, not through the nerves. The first hormone discovered was secretin, which was involved with digestion. Shortly thereafter adrenaline was identified, given that name because it was created by one of the adrenal glands (it's also called epinephrine). More and more hormones were steadily discovered. On Christmas Day in 1914, thyroxin—a hormone produced by the thyroid—was identified, and in the 1920s and '30s, testosterone, estrogen, and progesterone and their roles in regulating reproductive activity were discovered. Over time research showed that changes in the levels of these hormones could dramatically interfere with the normal reproductive cycles, ultimately leading to the creation of the birth control pill, which hit the market in 1957.

Two other adrenal gland hormones, cortisone and cortisol, were identified in the mid-1940s, and along with adrenaline, they were dubbed the stress hormones, because they all regulate levels of stress. Levels of adrenaline and cortisol were found to rapidly ramp up in response to perceived danger, key to the "fight or flight" response. In 1958, the isolation of another hormone, melatonin, was announced. This hormone was produced by the pineal gland, and in addition to affecting the pigmentation of skin, it played a vital role in regulating sleep patterns as well as the timing of reproductive cycles.

Research had also shown that rarely, if ever, does a hormone have a single effect on an organism. Most hormones affect a suite of different morphological and behavioral characteristics. Testosterone, for example, is involved not just in the development of the testis, but in aggressive behavior, as well as in the development of muscles, bone mass, body hair, and many other traits.

Dmitri had studied the literature on hormones and he knew that research had shown hormone production was somehow, though exactly how was not clear, regulated by genes. He thought the genes or combinations of genes that regulated hormone production might be responsible for many—maybe all—of the changes they were seeing in the tame foxes. The selection for tameness had triggered changes in the ways those genes were operating. Natural selection had stabilized the hormonal recipe for building a fox and its behavior in the wild. Now the selection for tameness that he and Lyudmila were imposing was *destabilizing* that formula.

Why, Dmitri wondered, would that be happening? The stabilization of an animal's behavior and physiology was suited specifically to its environment. Animals' mating seasons had been selected to coincide with the time of year when food and daylight were most favorable for the survival of young ones. Their coat coloring was optimized to camouflage them in their natural environment. Their production of stress hormones was optimized to cause them to either fight or flee from the dangers of their environment. But, what if they were suddenly transported to a radically different environment, one with different conditions for survival? That's what had been done with the foxes; their environment was now one in which being tame around humans was optimal. So the stabilization of their behavior and physiology that had been the result of natural selection in the wild was no longer the best formula, and adjustments had to be made. And Dmitri thought that under such pressure to change, the activity patterns of an animal's genes—the ways in which they regulated body functioning—might be dramatically altered. A cascade of changes might be unleashed. And it made sense that key among these would be regulatory, timing, changes in the production of the hormones that played such a vital role in optimizing an animal to its environment. Later he would come to add changes to the nervous system to his formula as well. He called the new process he was describing *destabilizing selection.*[5]

Lyudmila and the others needed time to process the idea. This theory was radical. The concept that the activity of genes could be altered without mutation being involved had barely started making its way into the literature. Dmitri was way ahead of the scientific community in making this conjecture that some of the changes in animals could come not from changes to DNA, but from genes already present being activated or deactivated in new ways. Up to this point, in conducting the experiment, they had been flying blind in a scientific sense, operating without a true theory. Now they had one. They had no proof for it yet, but it was an intriguing idea that might explain much if correct, and hopefully, Dmitri thought, in time the fox experiment would allow them to test this idea.

BELYAEV SECURED A GOOD PLOT OF LAND cut out of a lovely pine, birch, and aspen forest just four miles northeast of the Institute, and supervised the building of the fox farm. It was a barebones operation. Five wooden sheds were built, which could each hold fifty large pens. Feeding was done with a pulley system that allowed the workers to move large buckets of food up and down the sheds. A 100 square foot fenced-in playground where the foxes were allowed out to run and play for a period each day sat behind each shed. Fifty-foot-tall wooden observation towers were soon built to allow Lyudmila to sit and watch them through her binoculars, recording how they played and interacted with one another without disturbing them. There was a veterinary clinic as well, so any sick or injured foxes could be cared for right away.

In late fall of 1967, Lyudmila arranged for the transport of fifty female and twenty male foxes from Lesnoi to the new experimental farm. More would follow, until 140 tame foxes, of which 5%–10% were elites, were shipped from Lesnoi. Lyudmila worked with a farm manager to hire a small team of caretakers for the foxes, who would feed them twice a day and let them out in the yards for play. She took great care in hiring workers because she wanted to be sure not only that they weren't afraid of the foxes, but that they enjoyed being with them and would take very good care of them. She would find that these workers not only zealously cared for the foxes, many of them fell in love with them.

Most of the caretakers were local women, from the nearby town of Kainskaya Zaimka, and Dmitri arranged for a bus to pick them up and bring them back home each day. He was sure to chat with them a little when he visited the farm, as he did whenever he could spare a moment, which was not nearly as often as he wished. He was keen to meet these workers and he would walk up to them to introduce himself, seeking to shake their hands. One woman worker recounts that when she demurred, because she was embarrassed by how rough her hands were, making the excuse that they were too dirty, he took her hands in his and said to her, "Working peoples' hands are never dirty."[6] She was struck that a man of such high standing, the head of a major scientific institute, would treat her with such warmth.

These workers quickly developed a great fondness for the foxes. They watched zealously over them and went well beyond the call of duty to care for them, saving the lives of many pups who might otherwise have frozen to death, by keeping such a close eye on them. Sometimes fox mothers would neglect their pups right after birth, leaving them exposed to the frigid early spring weather. Even in April, the temperature could drop well below zero. The women would take off their thick fur hats and cradle the helpless little balls of fur in them, or hold them under their shirts in their bosoms until they had warmed up and had begun wriggling around.

On rare occasions, if someone was visiting the farm, the workers would pet the tame foxes and pick them up to show a guest how docile they were. The tamest foxes, even as fully matured adults, would allow workers to cradle them in their arms, or hug them tight, which felt good in the bitter cold of the Siberian winter. Some foxes would wriggle around in their arms when they held them, but others stayed so calm they seemed to be almost mesmerized.

A few foxes would lick the caretakers' hands when they reached into their cages on their daily rounds. But the workers didn't instigate this behavior. They were given a strict rule to remain as objective as possible with all of the foxes, no matter how irresistible they were, or how loudly the foxes protested for attention. Sometimes that was a challenge as the tamest foxes would whine and cry, causing a great clamor when the women entered the sheds, as if competing for attention, calling out, "Don't bother with her, come and see me!"

These tame foxes were establishing a strong sense of connection with the workers, and with Lyudmila and her research assistants as well. They would even allow people to look directly into their eyes, and they seemed to be looking back. With wild animals, included canids, staring directly at another member of a group is often taken as a challenge that leads to aggression. For a human to do so is to invite attack. But in domesticated species, as in many dogs, gazing into the eyes of humans is common.[7] Now these tame foxes were doing it too.

Though the caretakers resisted petting the foxes, they did begin talking to them quite a bit, always addressing them by their names, which were written on pieces of wood hung above their cages. Some of the workers chatted with them almost constantly while moving through the sheds during feeding time or letting them out to the yard for their playtime. They became more and more devoted to the foxes and engaged in the work being done with them. Starting with the first litter born on the farm, the women began assisting Lyudmila in naming the pups, which could be quite a challenge, requiring coming up with yet another six or seven names for each litter that all started with the first letter of the mother's name. They became Lyudmila's trusted eyes and ears, alerting her right away if a pup wasn't eating, seemed to have a cold, was scratching itself too often, or just didn't seem to be itself. Many of them regularly worked longer hours than their shifts, never complaining. Most of them loved spending as much time as possible with the foxes.

So did Lyudmila. She always had a great deal of data analysis and writing up of results to do, so she started her days by heading to the Institute of Cytology and Genetics to do some of that. If Dmitri was available, she'd also check in with him about the latest with the foxes and the work she had planned. But then she could head over to the farm for her favorite part of the day. Her first stop was usually to the veterinarian's office, to check on any problems there might be with any of the foxes. Then she would check in with the workers, whom she now thought of more as caretakers, and begin making her rounds through the fox sheds, always greeted by a great ruckus as the foxes jumped to the front of their cages, many of them now whining for attention, intently following her progress as she made her way from cage to cage. With the foxes so close by now, Lyudmila would also find herself heading over to the farm at off-hours, especially when she felt the need for an emotional pick-me-up. "I would take to the farm," she recalls, and "communicate with the foxes."

Generally, she devoted three to four hours each day to the foxes. A good part of that was taken up by collecting her standard housekeeping data: behavior, size, and growth rates, fur color, general body shape, and for the pups, milestones like when they opened their eyes for the first time. She also took daily notes on the behavior of the foxes toward her, her assistants, and the workers: and when it came to the young pups, how they behaved toward each other, who licked a human hand, or who wagged their tails. While the "official" behavioral data that would determine who parented the next generation was taken once when an individual was a pup and once when it had matured, these day-to-day notes on the foxes' behavior were critical, as they gave Lyudmila and Dmitri a fine-scale, deeper appreciation of the changes that were occurring.

With the extra room at the farm, Lyudmila also began breeding a population of control foxes, which would allow her and Dmitri to make rigorous comparisons between behavior and physiology of these foxes and those foxes bred for tameness. An important component of this comparative work would be measuring the levels of hormones in the two populations, with a focus on the stress-related hormones, which Dmitri and Lyudmila were sure were involved in some way with the foxes becoming tamer. Lyudmila had been able to take the blood samples only on occasion at Lesnoi because sampling required the help of workers to hold the foxes while she and her assistants drew the blood. Now she could do so regularly. This difficult and time-consuming sampling would soon produce rich rewards.

Another bonus of the experimental farm was that Belyaev could finally get to know the foxes intimately too, and he visited the farm as often as he could, sometimes even if he could only sneak away from the Institute for a few minutes with them. He especially loved watching the pups play out in their yards, seeing for himself the striking differences in the behavior of the tame pups versus the control pups. When he arrived, Lyudmila would sometimes bring a set of the tamest pups out so that he could see how they would lick his hands or roll over for him to pet their bellies. He was so enamored of the tame pups, marveling at how dog-like they were becoming, that he began imitating them when telling people about them, the way he acted out stories during dinners at his home with the staff. One of the researchers at the Institute recalls that "when he was talking about his foxes, changes would come over Belyaev, to his manners, to the way he was speaking, he behaved like a tame fox, he resembled a tame fox." He'd curl up his wrists as if he were begging, smile and open his eyes as wide as possible mimicking the foxes' excited reactions. The staff immensely enjoyed this because it showed a new part of him, what an animal lover he was.

Belyaev would occasionally bring visitors to the farm to show them the foxes, such as higher-ups in the Soviet Academy of Science or government officials on trips to Akademgorodok, and these people were also invariably charmed by the tame foxes. Lyudmila vividly recalls one such visit in particular. "I remember late in the evening after all the workers went home, Belyaev brought a famous army general, General Lukov, to the fox farm. I was given a heads-up that he was coming and that I must wait for the renowned guest." Lukov was a formal man, with a military bearing hardened by years of service, including the horrors of the Soviet front in World War II. Yet, when Lyudmila opened a cage that housed one of the elite females, and the fox scampered right over to Lyudmila and laid down next to her, the General's dignified demeanor melted away. "Lukov was astonished," Lyudmila says. "He squatted near the fox and petted her head for a long time." There was no denying that the tamest foxes were having a powerful emotional effect on people. And though the study of this effect was not a central component of the *design* of the experiment, they realized that it was a significant *finding* and might help to explain how domestication first started.

The rapid emergence of such solicitous behavior in some of the tame foxes fit well with Dmitri's idea that the process of wolf domestication had been kicked off by the animals first becoming tamer. Now perhaps the experiment had produced an important clue about why the process would have escalated thereafter.

One long-standing idea about wolf domestication was that humans had adopted wolf pups, perhaps choosing ones that were especially cute, with the most juvenile facial and body features. But what if it were the wolves that initiated contact, not the humans? Naturally more adventurous when it comes to humans, tamer wolves might have begun making their way into human encampments to scavenge for food. Maybe, given that they're nocturnal, they snuck into campsites in the night as our early ancestors slept. Or perhaps they had learned to closely follow human hunting parties to scavenge for prey. It's easy to understand why wolves who were relatively comfortable with human presence–naturally semi-tame—would have done so. We were a much more reliable food source than the wild. But why had early human groups accepted the wolves into their inner sanctums? Wolves on their way to becoming dogs might well have helped with hunting and acted as sentinels, warning of approaching dangers. But there must have been earlier stages of their transition before they were performing these functions particularly well. If the process of the silver foxes' domestication really was mimicking that of wolf domestication, then perhaps these same lovable *solicitous behaviors* emerged early on in wolves also. And maybe that made them more appealing to our early ancestors.[8]

But what would have driven the emergence of these behavioral changes in the wolves? Lyudmila was actively selecting the tamest foxes for mating. Is it plausible to believe that early humans would have actively mated wolves in a similar way? Perhaps they wouldn't have needed to. Natural selection would likely have favored the wolves who had gained access to such a reliable, human-based food source. The wolves that were friendlier to humans might have found themselves living in close proximity with other such friendlier wolves who were hanging around humans, and they might have selected their own, semi-tame, kind as mates. That would have created the radically new selection pressure for tameness that the fox experiment was applying. And as Lyudmila and Belyaev were seeing with the foxes, this new selection pressure favoring tameness might have been enough to trigger the kinds of changes they were seeing in their tamest foxes. The process would have taken way longer than with Lyudmila's artificial selection—as, indeed, it's thought to have with wolves—but the same essential force might have been at play.

Dmitri and Lyudmila also realized that the early emergence of the endearing behaviors in their foxes might provide some important new perspective on the evolution of animal expressions, and perhaps even on the nature of animal emotions, which were hotly contested subjects at the time. The debate had raged for decades about whether animals feel anything like human emotion, and whether the behaviors of animals that appear to be expressions of emotion really are, or whether instead they're simply automatic reflexes.

Charles Darwin was so fascinated by animal emotions that he made an extensive study of the subject, which he summarized in his classic book *The Expression of the Emotions in Man and Animal*. Published in 1872, the book was beautifully illustrated with drawings of animals' expressions, which Darwin commissioned from a number of the leading animal illustrators of the day, such as of a cat arching its back and raising its tail to show affection, and a dog looking up in a submissive and affectionate pose.

Darwin thought many animals have rich emotional lives, and he argued that their emotions, and their thinking abilities, too, are on a continuum with those of humans. "The difference in mind between man and the higher animals, great as it is," he wrote in *The Descent of Man*, "is certainly one of degree and not of kind." Throughout *The Expression of the Emotions*, he showed great empathy toward animals and the intensity of emotion they can feel: "The appearance of dejection in young orangs and chimpanzees," he wrote," . . . is as plain and almost as pathetic as in the case of our children."[9] Many human expressions, too, Darwin argued, are instinctive. And to illustrate, he included a striking set of photographs of people displaying characteristic expressions, such as those of grief, surprise, and joy.

A SCHOOL OF ANIMAL BEHAVIOR RESEARCHERS who eventually followed in Darwin's footsteps documented an astonishing array of complex innate behaviors, including, but not limited to, emotional behaviors. So compelling was the increasing evidence they produced of how genetically programmed animal behavior seemed to be that the idea that much of animal behavior was shaped by natural selection became the paradigm.

Generations of intrepid animal behaviorists had followed Leonid Krushinsky and others' example of observing animals in the wild and headed out to forests, meadows, streams, and mountain ranges to conduct studies. Others began observing animals both in the wild and in captivity through inventive new techniques. Three men in particular—Konrad Lorenz, Karl von Frisch, and Nikolaas Tinbergen—did so much to advance the understanding of animal behavior that they were jointly awarded the Nobel Prize in Physiology or Medicine in 1973. They carried out this work primarily in the 1930s, '40s, and '50s, and their fascinating findings were discussed often at both biology and psychology meetings.

The argument for natural selection as the driving force shaping animal behavior was strong. Many of the behaviors that Lorenz, von Frisch, and Tinbergen had observed generally conferred clear survival advantages. One of the most amazing of the elaborate behaviors observed was discovered by von Frisch, and performed by honeybees. Ingenious experiments he conducted with them revealed that they send signals to one another about where sources of nectar and pollen can be found when they return to their hives from foraging and perform a "waggle" dance.

Tinbergen observed remarkably complex and standardized behaviors in stickleback fish when it came time for mating. He discovered that the male always digs out a little crater of sand, which is almost always about two inches wide and two inches deep, and then covers it with a sticky wad of algae he forms from pulling bits together from the surrounding water into the crater. He then swims through this wad of algae to make a tunnel. Perhaps most amazing of all of is that he then changes color, from his normal bluish green to white on its back and a bright red on its underside. This is the trigger for a female to come mate, and when a female approaches, he directs her into the tunnel. She swims in, lays her eggs and leaves, and then he swims in and fertilizes them.[10]

For his part, Konrad Lorenz had caused a stir with his finding that Greylag goslings would treat him as their mother, becoming so attached to him that they waddled around behind him if he brought them out to a yard and walked around. Lorenz had noted how closely and rigorously goslings bonded to their mothers in the wild, never straying off and associating with other adult birds or with goslings other than their siblings. He was curious about this bonding process, and he performed an experiment in which he divided a group of freshly laid Greylag eggs in two, with one set being brooded over by the mother goose and cared for by her when they hatched and the other set placed in an incubator and cared for by Lorenz after they hatched. Those he had cared for became attached to him in just the way they would normally attach to their mother.

With more work, he discovered that there was a limited window of time during which this attachment formed. Whatever the goslings were exposed to during that window, they would treat as a parent, even if that was an inanimate object like a rubber ball. He concluded that the bond was formed instinctively, and he called the process imprinting. During this critical period in animals' early development, their genetically determined behaviors could be dramatically altered by the conditions they were exposed to.[11]

What was fascinating about Lyudmila and Dmitri's fox experiment results in relation to this work was that neither imprinting nor natural selection were the driving force behind the new behaviors or those that were retained beyond childhood in the tame foxes. Artificial selection for tameness was the driver. Exactly how, they did not know. But they were confident that Belyaev's theory of destabilizing selection provided the answers about what was happening with the foxes. To prove it, they would have to gather a great deal more evidence.

The foxes would not let them down.

4

Dream

With the foxes moved into their roomier new home at the new farm, Lyudmila was delighted that she could give them exercise, and she had the caretakers let them out into the yards behind the sheds to run around for about half an hour every day. This gave Lyudmila a whole new category of observations to make—she could now watch them play.

When the pups were still small, from two to four months old, they were let out in small groups of three or four at a time, with no adults, so that things wouldn't get too raucous. Just as with pups in the wild, who play with each other constantly when they're not sleeping or feeding, the pups on the farm scampered exuberantly around chasing and pouncing on one another, nipping at each other's tails and ears, and faux fighting by rolling around and wrestling with one another. Animal behaviorists refer to this kind of frisky roughhousing among animals as social play.

Many animals also engage in lots of play with inanimate objects, called object play, such as birds playing with twigs or shiny pieces of glass; cheetah cubs on the plains of the Serengeti who pat, carry, bite, or kick everything from bones to glass bottles; and dolphins playing with the air bubble rings they create. The tame fox pups threw themselves into this as well. Lyudmila bought them rubber balls, which they especially loved playing with, pushing them around with their snouts and jumping on them, but they had fun with anything they could get their little paws or mouths on, rocks, twigs, and some balloons also put out in the yards. As they got bigger and their jaws could open wide enough, they would pick the balls up in their mouths and race around the yard with them, trying to keep their prizes away from their brothers and sisters. This merging of social play with other pups and object play is also common in young animals, and is thought to help them develop the skills to keep a hold on their prizes from foraging or hunting that others in their own group would love to snatch away.

The adult foxes also played, and some of this was expected. Mothers in the wild play with their pups, and once in a while, Lyudmila observed them doing so and it pleased her greatly. Though social play between adults was rare in the elite foxes, they did engage in lots of object play with the balls and tin cans, and this was a big surprise. In the wild, adult foxes are preoccupied at almost all times with looking for food and avoiding predators. If an adult fox comes upon a novel object in the wild, it might smell the strange thing, or even paw it a little, trying to figure out what it is and whether it can be eaten. But that sort of exploratory behavior is quite different from what animal behaviorists categorize as object play, which continues after the animal has become familiar with an object and knows it's not food.

This avid object play by the adult tame foxes was another way in which they were acting like pups longer, and also more like dogs, who love social and object play as both puppies and adults. Seeing the foxes out in the yard from a distance, one would have assumed they were some kind of smaller breed of huskies.

Lyudmila and the assistants from the Institute now working with her would often go into the yards to observe the fox pups playing up close, but they never attempted to interact with them at all, and they were careful never to interfere with their pups' roughhousing.

But some of the tame fox pups took the initiave and began involving Lyudmila and the assistants in their play routine, running up close to them with wagging tails, racing around them, and hiding behind their legs or nipping their shoes and bashfully running away. They seemed curious, and excited, about these tall beings in their midst.

Lyudmila had expected that observing the foxes play would be an important part of her work. The ways in which animals play had long been studied. Ornithologists had observed many types of bird play, such as when they hang upside down from tree branches and swing back and forth with apparent glee. Chimpanzees had been observed playing and chasing each other in ways that looked much like children playing tag. Even some insects had been observed to play. In 1929, August Forel, an influential ant researcher, wrote in his book *The Social World of the Ants as Compared to Man*: “On fine, calm days when they are feeling no hunger or any other cause of anxiety, certain ants entertain themselves with sham fights, without doing each other any harm; but these games come to an end directly [if] they are scared. This is one of their most amusing habits.”[1] Today experts believe that these mock fights prepare ants for combat and courtship contests, which are key events in their lives.

Some observations suggest that animal play is, at times, a matter of pure enjoyment. Ravens in Alaska, northern Canada, and Russia are known to slide down steep, snow-covered roofs. When they reach the bottom, they walk or fly back to the top, and repeat the process over and over again. In Maine, ravens have been observed tumbling down small mounds of snow, sometimes while holding sticks between their talons. Chimps living on the Mahale Mountains of Tanzania do something remarkably similar, and again for no discernable reason. Videotapes have captured them stopping as they walk down a mount, marching backwards and pulling a handful of leaves as they proceed. They then often stop and somersault through the pile in apparent joy.[2] They seem to simply enjoy this play.

But play is also serious business, with many animal behaviorists arguing that it is essential for developing a host of social, physical, and psychological skills, preparing young animals for the challenges they will face as adults. Much social play is now thought to facilitate cooperation in groups of animals, such as when they're hunting or defending themselves from predators, and also to teach the young where they are in the pecking order and who they'd likely beat in a fight versus who they'd better watch out for.

Parents often lead their young in play, as when older meerkats teach the young how to hunt.[3] Juvenile kangaroos, called joeys, start play-fighting as soon as they leave their mother's pouch, often seen sparring with their mothers. Their stylized boxing isn't dangerous. Older partners self-handicap when they play with joeys by standing flat-booted and pawing, instead of throwing hard punches, teaching their little ones the finer art of sparring for when they will need it as they grow up, but sparing them the bumps and bruises.

Young ravens in nature manipulate and play with every object they encounter—leaves, twigs, pebbles, bottle caps, sea shells, glass fragments, and inedible berries—just as Lyudmila observed with the fox pups. Experiments in which Bernd Heinrich placed novel objects both in the field and in large aviaries show that this sort of object play by young ravens teaches them what is safe to eat when they are out foraging on their own as adults.[4]

For the most part, as with wild foxes, animals engage in less play as they mature. This was why the discovery that the tamer foxes were continuing to engage in object play as they grew up was important. Another youthful behavior was being extended into fox adulthood, as with the whimpering, hand licking, and general calmness. Lyudmila and Dmitri had more strong evidence for Dmitri's destabilizing theory that when you radically change selection pressures by choosing the tamest animals, you shake up everything, and a whole suite of changes follow.

IN THE TENTH GENERATION OF PUPS, born in 1969, two more striking physical changes occurred. One of them showed up in a precious little tame female pup. She had remarkable ears.

In wild populations, in the control population, and thus far in the experimental population as well, the ears of a fox pup are floppy until it is about two weeks old, and then they straighten out. This pup's ears, though, didn't straighten then, and they still hadn't straightened by the third week, and then the fourth, and fifth, and on and on. Her floppy ears made this little pup look almost exactly like a dog puppy. She was given the name Mechta, which translates to "dream."

Lyudmila knew Dmitri would be delighted by Mechta's ears, and she wanted to surprise him and let him discover Mechta for himself. But he was extremely busy that spring and didn't visit the farm until three months after Mechta's birth. To Lyudmila's delight, Mechta's ears were still floppy. When Belyaev saw her he exclaimed, "And what kind of wonder is this?!" He began showing a slide of her at all of his talks, and Mechta became something of a celebrity in the world of Soviet animal studies. At one conference in Moscow, after he flashed a slide of Mechta, a former classmate of Lyudmila's approached her and said, only half-jokingly, "So your boss is deceiving the audience showing us a dog pup, and portraying it as a fox!"[5]

The other new trait to appear in the tenth generation showed up in one male pup. This youngster displayed a new kind of piebald coloration. While in the prior generation, patches of white and brown had appeared on the belly, tail, and paws of a few tame pups, with this pup, a small white star patch appeared smack dab in the middle of his forehead.[6] This is another common feature of domesticated animals, seen especially in dogs, horses and cows. "We would joke," Lyudmila fondly recalls, "that [now] that a star had lit, this will bring us success."

So many of both the behavioral and physical traits of domestication had now shown up in the foxes that it seemed clear the experiment was working. But in order to prove Dmitri's theory of what was happening to the foxes, Lyudmila and he would have to find evidence that genetic changes were driving the process. They had no real doubt about this; in many cases the new traits had been transmitted from parents to offspring. But the science of genetics requires even stronger proof. So he and Lyudmila would need more.

The leading method at the time for establishing a genetic link to the appearance of traits was pedigree analysis, which involves comparison of traits through many generations of parents and offspring. Some variation in behavior and morphology always occurs between individuals of a species. No two foxes ever look or behave exactly the same. To conclude that the changes they had documented were truly tied to genes, the pedigree analysis would need to show the characteristic patterns of inheritance of traits that had been worked out over many years of research.

This sort of work was pioneered by the monk Gregor Mendel, who in the middle of the nineteenth century, tracked patterns of change in the colors of peas over many generations. Subsequent researchers had honed the methods of pedigree analysis to allow for considering a wide variety of traits. Lyudmila had drawn up family trees for all of the foxes, and her meticulous notes about all of their behaviors and physical traits allowed her to perform this analysis. It was arduous work, but she dug in, and the results were clear; much of the variance that they saw in the new traits in the tame foxes was the result of underlying genetic variance.[7]

Another way to garner powerful supporting evidence would be to replicate the results of the fox experiment with another species. In 1969, Belyaev decided to launch such an experiment. For this he turned to a young man named Pavel Borodin, who was in his last year as a biology major at nearby Novosibirsk State University. Pavel was friendly with Dmitri's son Nikolai, and when Dmitri saw him one day on a visit to the university, he asked what Pavel was doing as his senior year project. "He didn't detect any enthusiasm in my response," Pavel recalls, "and then Belyaev said: 'I'm not going to try to lure you away . . . you decide. But let's take a trip to the fox farm and you can take a look at what we're doing there.'" Borodin was excited at the prospect, and once he got there, he was hooked, amazed by how domesticated—and truly friendly—the foxes were.

Belyaev wanted Pavel to follow the same basic procedure they followed with the foxes, but now with wild rats, and he wanted him to select and breed not only a line of rats that were calm and tame toward humans, but also an aggressive line. That would allow for important comparisons between their offspring over time. Pavel was given lab space at the Institute of Cytology and Genetics, but as for the initial population of rats, he would have to go out and catch them himself. "The main source of my animals," Pavel recalls, "were the pig houses on farms. There were quite a lot of rats there. It was not easy to trap them because they were just clever animals, but anyway, I succeeded." After a few weeks of trapping, he had brought a hundred rats back to the lab.

Slightly modifying the technique that Lyudmila had developed for the work with foxes, Pavel would place a gloved hand into a cage and note whether a rat came over curiously to smell the hand or perhaps allowed him to touch it, or even pick it up. Some did. Others attacked, which was quite unnerving at first. But Pavel persevered, and after five generations, two dramatically different lines of rats developed, one increasingly tame, allowing him to pick them up and even pet them, and the other fiercely aggressive. Though Pavel moved on to other work thereafter, Belyaev decided to continue the experiment and hoped it would go on to produce more supporting evidence, which it has.[8]

Another important step in producing definitive genetic results was to begin also breeding a line of aggressive foxes. As it had with the rats, reversing the procedure that had been followed with the tame foxes and selecting them for aggressiveness toward humans should produce an increasingly aggressive group of animals, which would allow them to begin making rigorous comparisons between the three populations—the tame, the control, and the aggressive. The work breeding the aggressive line got underway in 1970.

While the elite foxes were a joy to work with, interacting with the aggressive foxes was not something the caretakers relished. The most aggressive were truly menacing, and they often lunged at Lyudmila with bared teeth when she tested them for selection. Those teeth are very sharp, and when foxes bite they bite hard. Most of the workers and scientists who helped Lyudmila with the foxes were terrified of these animals. "I looked at one aggressive fox," recalled one of them about a particularly upsetting encounter, "and she gazed straight into my eyes but didn't move . . . her fox eyes intently followed my every movement . . . I slowly brought my palm nearer to front side of the cage . . . and she reacted immediately. She threw herself to front side of the cage . . . her front paws against wire mesh . . . She had a really dreadful look: her mouth was open wide, ears were pressed tight to head and blind fury burnt in protruding eyes . . . When I looked into her eyes, I felt fear. My heart pounded rapid and blood rushed toward my head . . . I believed that she would have sunken her teeth into my face or neck if there were not wire mesh."[9]

Thankfully, one of the workers, a petite young woman named Svetlana Velker, was willing to take on the job. She was "a young, seemingly fragile woman," as Lyudmila tells it: "everyone was afraid to work with aggressive foxes, [but] Svetlana was astonishing everyone with her courage." Svetlana decided to just lay it on the line with the aggressive foxes. Tell them how it was going to work. "When she needed to handle . . . an aggressive fox," Lyudmila continues, "Svetlana told the foxes 'you are afraid of me, and I am afraid of you, but why do I, a human, have to be afraid of you, fox, more than you are afraid of me?'" Then she got down to business. "Belyaev always admired her bravery," Lyudmila recalls, "and used to say that they should raise her salary for this kind of work with aggressive foxes."

Others who would follow in Svetlana's footsteps had their own special ways of dealing with the aggressive foxes. Rather than the stern disciplinarian approach Svetlana adopted, Natasha, who works with these nasty foxes to this day, decided that these animals were what they were by no fault of their own. They needed to be loved just as much as the tame foxes. And that is what she did and continues to do. "I like my aggressive foxes most of all," Natasha says. "They are my children. I like the domesticated foxes, but I love aggressive foxes."[10] Whenever Lyudmila hears Natasha express this love, she laughs: "it is very, very rare" is all she can say. The courage of these assistants was to prove immensely valuable as the experiment progressed and the aggressive foxes allowed for important comparisons with the tame foxes.

In the meantime Lyudmila and Dmitri got started on an important comparison between the population of control foxes and the tame foxes. Belyaev had theorized that genetically linked changes in the production of hormones involved in the regulation of the reproductive cycle, temperament, and physical features were responsible for the appearance of many traits associated with domestication. In order to prove this part of his theory, they would have to measure levels of hormones in the tame foxes versus the control group. With the sophisticated equipment available at the Institute for doing this, Lyudmila could begin performing this analysis.

She decided to start by measuring the stress hormone levels of pups, to see if the tame pups had lower levels after the period when foxes normally start to become more anxious and fearful, between two to four months of age. This required a delicate procedure of taking blood samples from all of the pups, and this would have to be performed as quickly as possible, within no more than five minutes. Otherwise the levels of hormones would likely be elevated due to stress caused by the procedure, distorting the results.

Measuring hormone levels was a technical kind of work that Lyudmila had no experience with, so she sought the assistance of one of her colleagues at the Institute, Irena Oskina, who specialized in the work. But the problem was that Irena had never worked with foxes. So Lyudmila asked the caretakers, with whom the pups were so comfortable, to help her. They had to take samples at several stages of the pups' development, beginning before they were two months old and were still living in pens with their mothers, up until they were adults. The workers came through brilliantly. They would slowly reach in to get pups, trying not to alarm the mothers, and it was a true testament to how tame the adult foxes had become that the mothers didn't react viciously when they did so. When it came to the control foxes, again the workers rose to the occasion—the control fox mothers can be very vicious if they think their pups are in danger. Wearing the two-inch thick protective gloves that Lyudmila ordered for them, they managed with some practice to do the work with great efficiency.

When Lyudmila received the results of the analysis of the samples from Irena, she was delighted by the stark contrast in stress hormone levels. As expected, the levels rose in all of the foxes as they matured, but in the elite pups, the burst happened much later and the spike was much less pronounced, plateauing in adults at typically a 50% lower level than that of the control foxes. This was powerful confirmation of Dmitri's destabilizing theory regarding changes in hormone production.

AS BELYAEV BEGAN GIVING TALKS ABOUT all of these new results, the world scientific community became increasingly interested in the foxes. Dmitri was again allowed by the Soviet authorities to attend the International Conference of Genetics, which was held in Tokyo in 1968. His Japanese hosts were so taken with him and his talk that they presented him with some exotic domesticated roosters as a farewell gift. Somehow he got the live roosters on to the plane back to Novosibirsk.

Dmitri also began submitting papers to international academic journals, and the first article to appear in English outside of the Soviet Union—entitled *Domestication in Animals*—came out in 1969. But, up to this point, scientific attention to the work was mostly limited to the genetics community; animal behavior researchers hadn't taken much note of it. That changed when Belyaev received an invitation to the International Ethological Conference to be held in Edinburgh, Scotland, in September 1971. The conference was an invitation-only affair, bringing together the top researchers in the world. Dmitri was the first Russian scientist to be invited. The invitation came straight from the conference organizer, Aubrey Manning, who was one of the leading animal behaviorists in Britain. Manning had made it one of his missions to give a more international flavor to that year's conference. He wanted to reach out beyond the usual crowd from Europe and the United States and give the meeting what he called "a kind of United Nations feel."[11]

Manning had heard about the fox experiment and found the work fascinating. He had conducted his own graduate research under the supervision of Tinbergen, and he was a specialist in studying the links between genes and behavior. He and his wife, the geneticist Margaret Bastock, had performed groundbreaking studies with fruit flies in the mid-1950s that were some of the first to link a gene specifically to a behavior in an animal. Manning thought the strong evidence for genetic causes of behavioral change that the fox experiment had produced were very important for the animal behavior community to learn more about. When he wrote to Belyaev before 1971 asking if he'd speak, Manning's expectations were low. "This was at absolutely the height of the Cold War, of course, or at least the Cold War was pretty thick," Aubrey recalls, "and contacts with the Soviet Union were slim."[12] When Belyaev replied with an enthusiastic "yes," Manning was delighted that he had "extricated for the first time an ethologist from the USSR."

This was a big step for Dmitri and Lyudmila, and Lyudmila was excited about the opportunity to present their work to such an elite gathering. Manning asked Dmitri to bring a group of colleagues, and Lyudmila and a number of other researchers from the Institute were scheduled to attend. But shortly before they were due to leave, the government decided that only Dmitri would be allowed to travel. At least Lyudmila knew he would give a wonderful talk and their work would begin to make its way into the larger animal behavior discussion.

The venue for the conference was the David Hume Tower at the University of Edinburgh. Each day Belyaev, Manning, and the other attendees heard a suite of thirty-minute talks by some of the most respected animal behaviorists of the day,[13] including Tinbergen, who just two years later would be awarded his Nobel Prize as a cofounder of ethology. The sessions could be contentious, as there was a mini-battle going on in animal behavior between a European camp, who were primarily trained as biologists and tended to focus on genetics and study their animals in the field, and an American camp, who were primarily trained as psychologists and who focused on animal learning and worked with animals in the laboratory.[14] Some researchers in the latter camp had taken the arguments about conditioning to such an extreme that they denied that any animal behavior could be genetically "programmed," rather it was all the result of conditioning or learning. But a great deal of research by animal behaviorists out in the field was suggesting otherwise.

Some of the most important of the observations were made by biologist E. O. Wilson, who had traveled the world observing colonies of insects of many types. In January of the year of the conference, his landmark book *The Insect Societies* was published. He vividly portrayed rituals in insect colonies and included stunning photos and drawings of leaf cutter ants tending their elaborate fungus gardens, fertilizing their food source with manure they have gathered, or marching along with leaves many times larger their own bodies hoisted over their heads; of army ants returning to their nest with the remains of their scorpion booty; and of wasps applying a concoction of ant repellent to their nests. He described how in the colonies of some ant species, workers serve as living honey pots to be tapped by the community in time of need. They store nectar and honeydew in their guts and hang upside down from rocks in the nest. When drought conditions hit, the others turn on these living spigots for shots of energy. He also described the frightening opposite side of ant behavior, vividly depicting their brutal tactics in warfare, as when three ants will hold another in place while an attacker cleaves its body in half.

How could an animal like ants perform such purposeful feats, with such wide-ranging motivations? Much of it must be based on instinct.

Yet the behaviorists had produced strong evidence of animal learning. American psychologist Edward Thorndike had tested how quickly cats and dogs could escape from "puzzle boxes" he had built. He observed that they initially tried all sorts of ways to get out, and then, when by chance they discovered a route, they readily learned to repeat the process, getting out of the box faster and faster. This, he argued, showed that animals learned their behaviors by being rewarded for them, whether that was a certain way to approach a bird they were about to pounce on or licking a hand that then rewarded them.

Many animal behaviorists were beginning to think that it was quite possible that both genes and learning were involved in many of animals' complicated social behaviors. It was not a one-or-the-other scenario—learning could be layered on top of genetic predispositions, and what's more, the ability to learn may itself have an underlying genetic component. Belyaev thought this sounded about right.

Dmitri soaked up every bit of debate on these matters at the meeting in Edinburgh, and he enjoyed the sessions immensely, even if the speakers sometimes presented their work a bit quickly for someone whose first language wasn't English. A good crowd showed up for his talk, *The Role of Hereditary Reorganization of Behavior in the Process of Domestication*. The title was tantalizing—hereditary reorganization of behavior? What was that about? And domestication of what? Had Russian scientists been producing noteworthy work now that Lysenko had been banished? What would this Russian be like?

Dmitri read a prepared speech in English, and Manning recalls that he impressed the crowd. They hadn't known quite what to expect of him, but they hadn't anticipated a man who was so dignified and assured. They also hadn't expected anything like Mechta and her droopy ears. The results of the experiment in just over a decade were incredible.

Manning was so taken with Dmitri that he invited him to his home for dinner that night and had him shepherded over from the beautiful sixteenth-century Edinburgh student dormitory where he was staying. Belyaev's English was good enough to read a speech, but fast-paced dinner conversation was another matter, so a translator joined the dinner party. Dmitri had been hoping that he might get such a chance for socializing, and he had brought along some traditional Russian gifts. Manning was touched when Dmitri presented him and his wife with some beautiful lacquer bowls. The Cold War had shut Russian scientists off from this kind of free and comfortable social interchange with their peers around the world, during which they engage in so much creative exchange of ideas, often leading to new avenues of exploration. That seemed a shame to Manning, sitting with this warm and intelligent man who was so interesting. They became friends, and as he had with Michael Lerner after meeting him at the International Genetics Congress in The Hague, Dmitri kept up a correspondence with Manning in the years ahead. Manning hoped that before too long, he could make a trip to Novosibirsk to see the remarkable fox-dogs for himself.

An important sign that the results of the fox experiment were being recognized by the scientific community in the West was that shortly after the meeting in Edinburgh, the directors at the *Encyclopedia Britannica* wrote to Belyaev to ask if he would contribute an essay on domestication for the forthcoming fifteenth edition, a massive revision, also known as *Britannica 3*, scheduled for publication in 1974. Dmitri was thrilled, and immediately began penning the essay, which appropriately enough appeared immediately after the entry "Dogs."[15]

The study of links between genes and animal behavior was gaining speed in the 1970s, and the fox experiment work was at the forefront of this new wave of research. The first academic journal in the subject area, *Behavior Genetics*, had been founded in 1970 along with the creation of the Behavior Genetics Association, and in 1972, the Russian-born geneticist Theodosius Dobzhansky, whose work Dmitri knew well and who had emigrated to the US, was elected its first president. Russian genetics was most assuredly making a comeback, and Dmitri was acting as one of its leading ambassadors. In 1973, he was again allowed to attend the International Congress of Genetics, which was held at the University of California at Berkeley.

The meeting at Berkeley was both a scientific and cultural smorgasbord like Belyaev had never experienced before. As for the science, the meeting featured symposia, involving the world's leading authorities, on everything from "Genetics and Hunger" to "Dilemma of Science and Morals" and, more in line with Dmitri's own work, "Developmental Genetics" and "Behavior Genetics."[16] Everyone who was anyone in genetics research was there, and Belyaev had the chance to meet some of the most famous geneticists of the day and discuss his ideas with them. In between sessions and at night, the crowd enjoyed the hippy trippy street life of the town. Berkeley was a main hub of the student protests that rocked the nation, the epicenter of the Free Speech Movement, and freedom of expression was on display in all its glory. Street vendors, musicians, and jugglers competed for attention with hippies handing out protest pamphlets against everything from the Vietnam War to the nuclear arms race. Dmitri soaked it all in with great fascination, and he fondly told friends of Berkeley, which other attendees described as having "middle-class American youth garbed in saffron [who] danced to the repetitive beat of their Hari-Krishna celebration."[17]

During the course of the meeting, he and others in the delegation of Soviet scientists who had been allowed to attend decided to approach the organizing committee for the International Genetics Congress with an idea. Belyaev's administrative experience at the Institute made him the perfect leader of their effort. They proposed that the next International Genetics Congress, set for 1978, take place in Moscow. The organizing committee was intrigued. They were always looking for ways to make the International Congress of Genetics even more international, and Moscow would certainly do that. The policy of détente between the US and its allies and the Soviet Union, which had been instituted by President Richard Nixon in the early 1970s, had also made it possible to hold such a gathering behind the Iron Curtain. Having the meeting in Moscow would expose many geneticists to a group of scientists, and a scientific literature, they knew too little about. The idealists on the committee also dreamt that such a meeting might have ramifications beyond science: that perhaps this sort of reaching out might in some small way cool down the Cold War. And the committee was very sympathetic to the idea that holding the meeting in Moscow would show the world that the evils of Lysenkoism were a thing of the past.[18]

It was an ambitious undertaking, but, yes, the committee told Belyaev and his delegation: if you want to host the 1978 meeting in Moscow, we approve. Immediately Belyaev donned yet another title: Secretary General of the XIV International Genetics Congress to be held in Moscow.

THE NEW EXPERIMENTAL FOX FARM HAD ALLOWED Dmitri and Lyudmila to accomplish so much in just a few years. Lyudmila was observing the foxes much more closely through the whole course of the year and she felt her already intense bond with them strengthening. Deep inside, she knew that something was different. The emotional changes, the depth of feeling that these foxes began to express, and that they inspired in her and the caretakers, and anyone who visited the farm, could not be ignored.

It wasn't just the scientist in Lyudmila that was amazed by the increasingly adorable animals, but the human being. And that, she realized, was an important finding of its own, and surely part of the story of how dogs became so strongly domesticated, so bonded to us and intensely loyal to "their people." What if, she thought, I switch gears, and instead of resisting the animals' increasing charms, allow myself to explore just how far I could take these animals down the road of emotional expressiveness toward humans?

For a long time now she had pondered the limitations of the careful scientific data she and her team collected; it could tell her only so much. If she really wanted to know just how much social and emotional depth these tame foxes were capable of, she would have to give one of them the opportunity to live in the rich social environment of a home, with humans as its closest companions. Like dogs live. If the foxes were to truly become like dogs, they would have to develop the signature loyalty to their people that dogs show. While there was no question that the elites foxes had become intensely enamored of human attention, as of yet, they didn't differentiate between people. They were equally happy to see all humans. Maybe that would change if a fox was actually living with her.

She made a bold proposal to Belyaev. There was a little house off one corner of the fox farm. She told him that she would like to move into that house with one of the elite foxes to see what bonds might develop. He loved the idea and right away got her the authorization to use the house.

Lyudmila wanted to choose the fox she would bring to live with her very carefully. She decided that she would select an especially affectionate female elite as the "Eve" for the experiment, to give birth to the fox she would move into the house with her. By this time many of the elite females were good candidates, but this was a unique experiment, and she wasn't about to rush into a choice. She pored over her notes and charts of data, evaluating the combined information about the elite females' stress hormone levels and behavior, electing a set of top candidates. She then went to their sheds and observed them closely, assessing them anew. After many days of this evaluation, she had her fox.

Her name was Kukla, which in Russian means "little doll." She was one of a handful of tame females who had become fertile (but not pregnant) twice during a year, and there was something especially beguiling about her. When Lyudmila approached her cage, Kukla would spring to life, emphatically wiggling her tail and squealing with what could only be described as sounds of pure delight. "She is asking for it," Lyudmila thought. The only problem was that Kukla was tiny for an adult female. She had been the runt of her litter, and Lyudmila wondered if she shouldn't select a fox of sturdier stock. In the end, she went with her gut. Kukla it was.

The father would be a fox named Tobik, a tame fox of the same generation as Kukla. They mated successfully and seven weeks later, on March 19, 1973, little Kukla gave birth to four healthy pups—two males and two females. As soon as they had fully opened their eyes Lyudmila went to see them. She found several of the caretakers crowded around them, doting over them like they were their own children and grandchildren.

Right away Lyudmila was drawn to one pup, who was such a puffy little wonder of fur that the workers had named her Pushinka, which translates to "tiny ball of fuzz." As Lyudmila kept observing her over the next few days, she saw that Pushinka was intensely solicitous of human attention. Pushinka was already creating such a powerful sense of connection with people that she seemed the perfect choice as Lyudmila's housemate. And, in this one case, because Pushinka was going to live with Lyudmila in a special experimental house, the workers knew it was okay to give in and play with her to their heart's delight.

During the next few weeks, as Pushinka grew stronger and more rambunctious, one of the caretakers, Uri Kyselev, grew especially fond of the adorable little pup, and he made a surprising request. He asked Lyudmila if he could take Pushinka home to live with him for a while, before Lyudmila moved her into the little house for the longer-term living experiment. Lyudmila thought about this and decided that it wouldn't interfere with her plan; in fact, this would allow her to see whether Pushinka would form a special bond with whatever person she was living closely with. Uri and Pushinka lived together at his house, just the two of them, from April 21, when she was one month old, until June 15, 1973. Pushinka adjusted wonderfully, giving Uri no problems. He even began taking her out for walks on a leash. He also found that he could let her out in the back yard, off leash, and she would come bounding right over to the door and come back inside when he whistled. This responsiveness to being called had never been seen with the foxes—quite the contrary. When on occasion a tame fox on the farm had broken free from a caretaker on the way to playtime or while being examined, the caretakers would call to them, but they'd never responded this way. Getting them back required much chasing around the farm, and a couple of them escaped the farm and ran away. Pushinka's behavior on this front was a great indication that Lyudmila had selected well and that her upcoming house experiment held more surprises in store.

With so many discoveries being made about Pushinka already, Lyudmila had decided that she would wait a little longer still to move

Pushinka into the experimental house so that she could observe how Pushinka would reintegrate into the farm fox society after living with Uri. Would she adjust to sharing her life with foxes again, or would the experience of living one on one with a human have changed her behavior with foxes? Wild animals brought into human society often have trouble integrating back into life with their own species. Lyudmila thought this was a good opportunity to observe how Pushinka would manage the transition and how the other foxes would respond to her. While she observed that Pushinka had no trouble interacting normally with the other foxes after she returned, she did demonstrate one striking change in her relations with them. If one of the other foxes was being aggressive toward her during playtime in the yard, as the fox pups often did with one another as they got older, Pushinka would seek the protection of the caretakers, hovering around their legs and keeping them between herself and the other fox. This was another first. Up to then, the foxes had managed their interactions with one another entirely between themselves.

Given that Lyudmila's primary aim of her planned house living experiment was to see how much like a dog Pushinka might learn to be from spending more time with people, she decided that it would be fine for the caretakers to take Pushinka for walks with them on a leash, as Uri had. Pushinka loved this. Knowing how obediently she had come for Uri when he called her, Lyudmila also allowed the caretakers to let her out without a leash, and she would follow them around as they did their feeding and cleaning.

Lyudmila now decided to update her plan for Pushinka again. Before long, with Pushinka closing in on one year old, it would be mating time, and Lyudmila decided to wait until Pushinka became pregnant before moving her into the experimental house. Then not only could she observe how Pushinka would adjust, but also whether her litter of pups would socialize differently.

On February 14, 1974, Pushinka mated with a tame male named Julsbar and, at long last, on March 28, 1974 she and Lyudmila moved into the little house. An unprecedented study in the history of animal behavior was about to begin.

5

Happy Family

Lyudmila's plan for living with Pushinka was to spend most of her days and nights at the little house with her, but so that she could also have some time with her human family, she arranged for her long-time assistant and friend, Tamara, along with a young graduate student, to help out by taking over some days and nights. Marina, Lyudmila's teenage daughter, and research assistants from the Institute would also occasionally work shifts if neither Tamara nor Lyudmila could be there. Whoever was on shift would make detailed journal entries throughout the day and evening about all aspects of Pushinka's behavior.

The first tense day in the house, when Pushinka paced around so anxiously and refused to eat, was unnerving for Lyudmila. Given how smoothly Pushinka had adjusted to living with Uri, Lyudmila had expected she'd have an easier transition. Maybe she was tense because she was pregnant? At least she'd comforted herself by sleeping for a while next to Lyudmila's daughter, Marina, and Marina's friend, who were there for move-in day. The next day, Pushinka was less agitated. After Lyudmila had stepped out for a moment, and then returned, Pushinka "met us at the door," Lyudmila jotted down, "like our dog," but her mood swung wildly from happy playfulness to listlessness, and she was still refusing to eat. All she had that day was a bit of raw egg. When Lyudmila offered her some chicken legs, one of her favorite snacks, Pushinka hid them in the corner of her room, a behavior familiar to dog owners. She wouldn't spend any time in her den and again hardly slept.

On the third day, Pushinka was still not eating or sleeping normally, and Lyudmila was getting quite worried. Pushinka paced restlessly around the house still spending no time in her den. She did seem to be taking comfort at Lyudmila's presence and she was increasingly seeking her attention. When Lyudmila sat down to work at the desk in her room, Pushinka came and lay down on the couch placed by the bed, finally getting some more rest.

Lyudmila was relieved and delighted when after yet another restless day of not eating, on the fourth night, Pushinka quietly jumped up on the bed while Lyudmila slept and curled up beside her. When Lyudmila woke, Pushinka scooched up closer to her head, placing her face right up next to Lyudmila's, and when Lyudmila put an arm under Pushinka's head, Pushinka rested her two front paws on it, cuddling like a child in its mother's arms. Finally, she seemed to feel at home.

But the following day Lyudmila was surprised to find that Pushinka was emotionally overwrought again, so much so that Lyudmila wrote in the journal that she seemed to be "at the edge of a nervous breakdown." It had now been five days, and she was still eating hardly anything at all. Lyudmila was now very concerned, and she called the farm's vet, who gave Pushinka a glucose and vitamin injection. Thinking that perhaps having her male mate with her would sooth Pushinka, Lyudmila had Julsbar brought to the house, but while he seemed delighted to see Pushinka, the feelings were not mutual. Pushinka screamed at him and chased him around the house, biting him several times. Lyudmila had him taken away immediately.

Dmitri was concerned by the news about how Pushinka was behaving, and he came to the house to check on her. Something about his presence seemed to calm her, and that day for the first time during the normal daytime rest period, she lay down at Lyudmila's feet as she worked at her desk and seemed content. That night, she finally began eating normally. The adjustment had been more traumatic than expected, but from that day on, Pushinka lived happily in the house, sleeping and eating normally and forming a stronger and stronger bond with Lyudmila.

Pushinka lay by Lyudmila's feet while she worked at her desk, and she loved for Lyudmila to play with her and take her for walks around the area. A favorite game was when Lyudmila would hide a treat in her pocket and Pushinka would try to snatch it out. Just as dog puppies love to do, Pushinka loved to playfully bite Lyudmila's hands, never hard enough to hurt. She also enjoyed lying on her back, paws in the air, inviting Lyudmila to pet her belly. She usually slept in her den, but some nights she would sneak up onto the bed with Lyudmila.

In the evenings, after an afternoon of rest, she was especially rambunctious and would pester Lyudmila to play with her, tossing around a ball on the floor, presenting her belly for a rub, or running over to her with a bone in her mouth. Out in the yard behind the house, Pushinka would sometimes take a ball in her mouth, trot over to a part of the yard that was elevated, release the ball and chase after it as it rolled down the slope. Again, and again, and again. When Lyudmila let her out in the yard, she'd always come bounding right back inside when Lyudmila called her. Just like a dog.

Pushinka gave birth on April 6, a day that Tamara was filling in for Lyudmila at the house. Just before her water broke, Pushinka came over to Tamara, and as Tamara was petting her, she delivered her first pup, right there. She cleaned her newborn and took it to her den, where she gave birth to five more pups. Lyudmila rushed to the house when Tamara called her with the word, and to her astonishment, when she arrived, Pushinka carried one of her pups in her mouth over to her, placing it gently at Lyudmila's feet. Normally fox mothers zealously guard their pups, and even the elite females were aggressive if the caretakers approached them right after birth. Lyudmila's own maternal instincts kicked in and she chastised Pushinka, saying "Shame on you! Your pup's going to get cold!" Lyudmila picked the pup up and brought it back to the den. Her face lit up with a smile as she thought just how extraordinary it was for Pushinka to do such a thing with her newborn.

The pups were all given names starting with a P, in honor of their mom: Prelest (Gorgeous), Pesna (Song), Plaksa (Crybaby), Palma (Palm Tree), Penka (Skin), and Pushok (the masculine version of fluff or "little fuzz," because he looked so much like his mother). When the pups finally opened their eyes, they were notably solicitous of human attention. Little Penka in particular was immediately affectionate, "cheerful to see humans, wiggling her tail with great excitement" at the sound of her voice, as Lyudmila wrote in the journal. Within two more weeks, all the pups were equally responsive to her voice, and they would come running out of their den when she entered the room.

With so much time to observe them closely, Lyudmila noted that the pups had distinctive behaviors. Prelest tended to dominate the others, being more aggressive in play with his siblings. Plaksa wasn't as fond of being petted as the others, while Pesna was downright stoic and often made an odd mumbling-like growl sound, as if talking to herself. Palma liked to jump onto tables, and Penka was especially fond of playing ball and was deemed a "sleepyhead" in Lyudmila's journal. Pushok craved interactions with Lyudmila more than all the others.

Lyudmila was especially drawn to the little tail-wiggling, sleepyheaded Penka, who was the smallest of the pups and was often picked on by her siblings. She tended to stay by herself, away from her siblings, and she was anxious around people in a way the others weren't, including with Lyudmila at the start, who wrote in the journal that Penka seemed to be contemplating whether "she should completely trust me." But before long, Penka clearly decided she could and her demeanor totally changed. Some days Penka would fall asleep only if Lyudmila picked her up and rocked her softly in her arms.

Lyudmila often played with Pushinka and the pups in the yard, throwing balls for them to tackle and push around. She would also run around so they could chase her. Penka would apply herself especially hard in this chasing game, jumping on Lyudmila's back when she stooped down and holding on in a kind of fox hug. When Lyudmila sat on the couch, Penka would jump up next to her and sniff her hair and ears and very lightly nip at her nose, cheeks, lips, and ears, which the other pups didn't do. Penka also made a sound unlike the vocalizations of the others, a sort of cooing noise that struck Lyudmila as a clear attempt to communicate with her. She often seemed to be wanting to tell Lyudmila something. "Penka is following me and talking all the time," Lyudmila wrote in the journal.

Penka appeared jealous of Lyudmila's attention to the others and she would sometimes lash out at them when they approached her if she was with Lyudmila, which, being so small, she rarely did otherwise. She would also seek Lyudmila's protection from the others. One day Penka found a cracker on the floor and was running around with it in her mouth, with her siblings in hot pursuit, and she jumped up on the couch next to Lyudmila and stashed the cracker behind Lyudmila's back. Then she defended her position against her siblings.

As one who had lived with dogs growing up, Lyudmila had seen this sort of behavior many times. But it was new in the foxes. As someone who understood animal behavior, she was acutely aware that she had to be careful when it came to attributing emotions and feelings to the foxes. Whether Penka was feeling anything like human jealousy would be impossible to say for sure. Dog experts know this conundrum about interpreting animal behavior well. In *For the Love of a Dog,* Patricia McConnell tells this story about one of her dogs, Tulip, who had discovered that a sheep that she had long played with had died: "[Tulip] sniffed Harriet's body, circling, sniffing and repeatedly nudging her. After a few minutes, she lay down beside the body. She placed her big, white muzzle on her paws, sighed once—a long, slow exhalation of what we'd call resignation in a human—and then refused to move . . . I don't remember how long Tulip lay beside Harriet, but she wouldn't leave her voluntarily . . . Tulip looked for all the world as if she knew that Harriet had died . . . There's only one catch. She behaves in similar ways to pigeons she's killed herself, and last week, to an ear of corn I'd given her to chew . . . There's a danger in attributing any emotion to a dog's behavior, because we are so often wrong about it. That doesn't mean they don't have emotions, it just means that we need to get better at reading their expressions."[1]

Along these lines, Alexandra Horowitz devised an ingenious experiment to study the "guilty looks" dogs proffer when being caught in the act: looks that Darwin described as eyes "turned askant," and others have portrayed as "plead[ing] forgiveness by frantically offering his paw," "slink[ing] back in a submissive way" or performing a "Tai Chi slink," often with "tail between her legs."[2]

Horowitz placed a scrumptious treat in a room and then had a dog's owner tell their pet that it was either ok to eat it or it wasn't (no!). The owner then left the room and the dog was left alone with the treat. The catch was that when the owner returned and the treat was gone, sometimes the dog was the culprit, but in other cases, unknown to owners, Horowitz had removed the treat. When owners castigated their dogs for the missing treat, they got a "guilty look" in return regardless of whether the dog was a treat stealer or not. It wasn't "guilt" at breaking the rules, dogs just don't like being scolded.[3]

So whether Penka was feeling jealousy for Lyudmila's attention, she couldn't know for sure. But what she could know was that this little pup was forming a special bond with her. It kept growing stronger as the pups matured, and Lyudmila felt the strength of the bond too. In time, Penka would need her special human friend to intervene and save her from increasingly rough treatment by her siblings as Pushinka allowed the pups to take full charge of adjudicating their own quarrels.

Pushinka was a good mother, and she played quite a bit with her pups, watching out for them in the early days. She loved to play pursuit games with them. Pushinka and the pups would chase after

Lyudmila in the yard, pulling at her clothing and nipping at her legs and feet. But as attentive as Pushinka was, when the pups' play got rougher as they matured, they had to fend for themselves, and little Penka often needed Lyudmila's protection. Pushok was particularly aggressive toward Penka, often tossing his sister "militant looks," as Lyudmila recorded, and following that up with an attack. Lyudmila couldn't always be there to protect Penka, and one time she was attacked so viciously that bits of her neck hair were ripped out. Lyudmila called the vet, and he took Penka to the clinic for care.

During her convalescence, which took place at the main part of the farm, where she could receive proper care, Lyudmila visited often, and Penka would perk up markedly when she arrived. When she left and Penka would whimper, Lyudmila was deeply moved, and she wrote about how hard it was for her in the journal. "I visited Penka at 6 pm," Lyudmila wrote, ". . . she came in when I called her. She greeted me quietly, without complaints . . . she immediately climbed on hands." Day after day this happened: "Penka was sitting sad and got happier," Lyudmila noted in her journal, only at her approach. Once she was with Penka, her little fox friend "would not leave [her] side . . . she ran to my feet, like a small puppy. Allowed me to do anything, then fell on her side when I started to pet her." How could Lyudmila not be moved?

As much as she loved Penka, Lyudmila had feelings for all the pups, and they, apparently, for her, and for her daughter, Marina, as well. "The pups are really crowding around me and Marina," Lyudmila wrote in her journal. "They climb on our lap 3–4 of them at a time . . . 'singing' something." It was hard for her to say more about this vocalization. It sure sounded like a sound that might indicate contentment, but vocalizations weren't her specialty, and again she paused to remember how difficult it was to assess emotions in animals; and so for the time being, she'd write it down in the journal and make a mental note to come back to these vocalizations one day in the future.

When the pups were in a more rambunctious mood, they would purposely run into Lyudmila and "wiggle their tails, and lay on the floor, panting." They were leading the carefree life. Lyudmila would write of walking into one of the rooms of the house and seeing all the pups, and "they were very funny sleeping, with no worries and with no fear."

A deep bond also developed between Pushinka and Lyudmila. As Pushinka's pups grew and she had to spend less time watching out for them, she turned her attention more toward Lyudmila again, constantly seeking her company. If Lyudmila was over at the other side of the back yard, Pushinka would come and stand by her, prompting her to play with her and pet her, lying at her feet and looking for her to scratch her neck. When Lyudmila had gone out to do some work at the Institute or for some time with her family, Pushinka would greet her at the door excitedly when she got back, wagging her tail.

Another dog-like behavior she developed was treating different strangers who came to the house as individuals, not just generic humans. She was quite friendly towards people in general, but just as dogs will sometimes bark hostilely towards certain people and be immediately affectionate towards others they've just met, Pushinka became more wary of some visitors than others. She had continued to hide special food Lyudmila gave her, like chicken legs, around the house, and one day when the cleaning lady came to the house, Pushinka rushed out of her den and went hurriedly around from corner to corner of the rooms gobbling up her goodies. She seemed to be leery of this woman who might take her precious treats. When one of the male researchers at the Institute, Anatoly, came to the house, she moved the pups out of the room, as though to protect them from him. When Pavel Borodin of the wild rat domestication experiment came over, as he did often, sometimes spending the night if Lyudmila had something pressing to attend to, Pushinka laid on her back in front of him looking for him to rub her belly. She seemed to have developed an understanding that those who actually lived in the house with her and the pups—not only Lyudmila, but the other researchers who spent days and nights there—were in a special category.

But it was with Lyudmila that she developed the strongest bond, like that between a dog and its master. She became increasingly protective of Lyudmila, and also acted jealous of her attention. When one day Lyudmila brought a new tame female fox named Rada to the house for a visit, Pushinka attacked her, pushing her out of the house and into the backyard. Pushinka also acted angry with Lyudmila: "I felt like Pushinka did not trust me anymore," Lyudmila noted. "She won't even let me pet her." But things got better fast. "Our interactions returned back to normal," Lyudmila wrote, "once I got Rada out of the house."

The intensity of the bond was clear, but even so, Lyudmila was shocked at the way Pushinka demonstrated her loyalty one night.

It was July 15th (1974), and Lyudmila was sitting on a bench outside the small house to relax for a while, with Pushinka resting at her feet, as she often did. The sound of footsteps approaching the nearby fence that surrounded the house roused Pushinka. Lyudmila assumed it was the farm's night guard on patrol and didn't give it a second thought. But Pushinka had other ideas. Lyudmila had never seen her react aggressively toward a human. But now she apparently sensed danger. She bolted off in the twilight toward the perceived intruder, and what Lyudmila heard then stunned her. Pushinka let out a series of barks. Aggressive foxes sometimes made short, menacing sounds at people who approached their cages. This was different. Pushinka hadn't been approached, she had gone in pursuit to bark at someone. She sounded just like a guard dog. Lyudmila immediately thought, dogs bark to protect their humans, foxes do not.

Lyudmila hurried over to the fence and saw that it was in fact the regular night guard who had spooked Pushinka, and as soon as Lyudmila started talking to the woman, Pushinka sensed all was well, and stopped barking.

To this day, Lyudmila has difficulty finding the right words to express the flood of emotions that overcame her when she heard her friend barking that July evening. She was overwhelmed with a sense of pride. As for Pushinka, she seemed plenty proud herself.

Lyudmila had been curious to discover whether living with a

human, or a group of humans, would elicit a loyalty to particular people in the elite foxes akin to that shown by dogs, and with Pushinka, there was no question that a depth of connection with and protective behavior toward Lyudmila had developed.

LYUDMILA AND ALL THOSE AT THE FARM STARTED referring to the house at the edge of the farm as "Pushinka's house." The days Lyudmila spent there were never boring. Pushinka's pups were becoming increasingly rambunctious and they began avidly playing games with Lyudmila. "If one of them jumps on my lap," she wrote, "the second one is pulling the first one away, a third one is pulling away the second one and so on." The pups would clamber up next to her if she sat on the couch, sniffing her head and licking her ears. They also loved a game of hunting that Lyudmila made up for them. She'd put a cloth or bathrobe down on the floor and move her hand under it like a mouse and they would rear up and energetically pounce.

The pups also began to engage in sibling rivalries, and Lyudmila, as a second mom, sometimes had to settle everyone down. "Pushok was chasing Penka in the shed," she wrote "[and] by the time I got there, Penka seemed like she was fed up with that, and she let me pick her up, very happy when I took her back in the house."

By age nine months, Pushinka's bunch were no longer pups, they were getting close to mating age, and Lyudmila and the team had to make some decisions. They couldn't possibly keep Pushinka, and all her kids and grandkids, in the tiny experimental house. They decided that they had to select only a few pups born each year to Pushinka or her children for living in the house and the others would live with the rest of the elite foxes on the farm. They continued to give each new pup selected a "P" name, in honor of Pushinka, and so in short time the brood included Proshka, Pamir, Pashka, Piva, Pusya, Prokhor, Polyus, Purga, Polkan, and Pion. As they grew up, each displayed some distinctive qualities. Proshka especially liked to sniff Lyudmila's hair. Polkan spent her days following Lyudmila wherever she went. Proshka's "favorite job," as noted in the journal, was to chew on Lyudmila's shoes. Pamir was especially "talkative," making a chatting sound to himself, and Pirat showed more independence than the others.

Much as she enjoyed her time at the house with all of them, Lyudmila also decided that she would no longer sleep over at the house most nights, and she started heading home to spend time in the evenings with her family. The foxes hated to see her go and would follow her to the door, and she felt guilty leaving them at first. On the positive side, as she approached the house each morning, the foxes would be eagerly peering out the windows and they would greet her at the door with a great burst of excitement.

By the start of 1977, Pushinka's house had fallen into sore disrepair, so Dmitri secured funds to build a new home in order to keep that part of the experiment alive. He and Lyudmila decided to use the occasion to make a key change in how she would conduct the house study. Lyudmila needed more time to work on analyzing all of the voluminous data the experiment on the farm was producing about the changes in the foxes, so they decided that she would observe Pushinka and her clan for fewer hours each day. The new house would be divided into a section for the foxes and another part for Lyudmila to work in, which would be closed off from them so that she could work peacefully. She would spend at least a couple of hours each day with the foxes in their part of the house and out in the yard.

When Pushinka and two of her daughters, along with two of her grandchildren, were moved into the new house, they were not happy with this arrangement. They seemed to miss unimpeded access to Lyudmila, and Lyudmila also found herself missing their company. Pushinka took the long daily separation from Lyudmila especially hard, and she regularly tried to sneak into Lyudmila's part of the house when Lyudmila came to see the foxes. Sometimes she was successful, wriggling by Lyudmila, and when Lyudmila made her go back to the fox side of the house, Pushinka would loudly vocalize her displeasure. Lyudmila noted that Pushinka seemed to remember the old life in the prior house, writing, "When Pushinka was in the yard, she used to look at the old house, where she happily lived side by side with people."

Seeing Pushinka unhappy was difficult for Lyudmila, and sometimes she bent the rules. "Pushinka was unusually sad and affectionate [today]," she wrote in a journal entry. "She put her head on my feet and lay there for a long time, looking up into my eyes with sadness and devotion." That day Lyudmila allowed Pushinka to spend some time with her, investigating the human quarters. No one wants to see their friend that down.

Perhaps because they had less time with Lyudmila and the assistants who helped her in the study, the foxes became more defensive about the time they did have with their humans. They would rush towards anyone who came over to their side of the house and compete for attention. Normally, they played with one another perfectly well and generally they enjoyed one another's company. But when Lyudmila or Tamara, the assistant who spent the most time at the house, would sit down to relax, if they petted or paid special attention of any kind to one of the foxes, and another tried to join the party, the invading party would be warned off by an aggressive growl.

The foxes at the house also became more protective of Lyudmila and "their" usual people. One day in July 1977, a researcher and a student from the Institute who had never been to the house stopped by to see the foxes. When they entered the house, Pushinka became furious. The only other time Lyudmila had seen her react so aggressively was the night she had chased after the guard and barked at the woman. Lyudmila had never heard her bark in quite that way again, and she wasn't barking now either, but she was growling very aggressively, which was a behavior the elite foxes didn't normally exhibit. Pushinka was clearly differentiating people who were associated with the house from strangers. There seemed no doubt that Pushinka was learning some of her new behaviors.

THE DEBATE ABOUT THE RELATIVE IMPORTANCE of innate versus learned behavior, which Belyaev had soaked up at the Edinburgh conference in 1971, had not died down in the years since. Lyudmila's findings with Pushinka offered powerful evidence for the view that a hard line one way or the other on this issue was simply misguided.

Particularly vehement controversy had erupted over the work of primatologist Jane Goodall, who made astonishing observations of chimpanzees at the Gombe Reserve in Tanzania, on the east coast of Africa. She had started observing them in 1960, at the suggestion of paleontologist Louis Leakey. Leakey and his wife, Mary, had been making remarkable finds of the fossilized skeletons of proto-human ancestors at Tanzania's Olduvai Gorge, and Louis thought observing the behavior of primates might help to illuminate how those early human ancestors had lived. Goodall's reports on the nature of chimp society, and how human-like so much of their behavior was, had captivated the public from early on. In some animal behavior circles, some people strenuously objected to many of Goodall's assertions about the implications of the behavior she had seen. In her book *In the Shadow of Man*, she wrote captivating descriptions of the close-knit nature of chimp communities: "I saw one female, newly arrived in a group, hurry up to a big male and hold her hand toward him. Almost regally he reached out, clasped her hand in his, drew it toward him, and kissed it with his lips. I saw two adult males embrace each other in greeting." The young chimps seemed to revel in their daily camaraderie with "wild games through the treetops, chasing around after each other or jumping again and again, one after the other, from a branch to a springy bough below."[4]

Goodall argued that individuals in groups displayed distinctive personalities, and that while mother-child bonds were the most powerful, strong social ties bound not only members of immediate families, but also larger groupings. Chimps seemed to genuinely care about members of their groups. They shared food, and came to the aid of one another when necessary. To her horror, as she continued to observe the chimps in the mid-1970s, she also observed acts of extreme violence, watching more dominant females kill the offspring of other females in a group, as well as group killings by males, which sometimes even ended in them eating the group member they had killed. That animals would kill one of their own in such a strategic manner had also been considered a uniquely human characteristic. It wasn't, and that was disappointing to Goodall. "When I first started at Gombe," she wrote many years later, "I thought the chimps were nicer than we are. But time has revealed that they are not. They can be just as awful."[5]

The seemingly human-like behavior of the chimps suggested to Goodall, and many others, that they had higher order thinking abilities, and more human-like emotions, than primatologists had thought. This was fueling new speculation about the nature of animal minds and how sophisticated some animal thinking, and learning, might be. The work also stirred up new ideas about how much more like our primate ancestors we humans might still be. But some ethologists thought Goodall had gone way too far in her conjectures about the chimp mind. They argued that she was anthropomorphizing, projecting human qualities onto the chimps that they didn't really have. The fact that she had given names to the chimps, like Greybeard, Goliath, and Humphrey, fueled that fire. But especially strong objection was made about her assertion that chimps were so smart they'd learned to become tool makers. Among her earliest observations, she watched chimps strip the bark from slender twigs and then insert them into termite mounds, pulling them out and slurping up a teeming feast. This seemed to her clear evidence of tool use, which no primate but humans had been thought capable of. Some animal cognition experts were unconvinced; they argued that this behavior couldn't be taken as evidence of anything like human-style problem solving or reasoning.

Certainly the learning Lyudmila was seeing in the foxes was nothing on the order of that required for tool use, but Dmitri and Lyudmila considered it important in understanding the process of domestication. They were not specialists in animal cognition or emotion, and they weren't equipped to perform a study of the foxes' cognitive abilities, or any analysis of whether they felt anything like human happiness and affection when they were wagging their tails, whimpering, licking hands, or rolling over onto their backs. Gaining definitive insight into animal emotion, they believed, might not be possible, which many experts today argue is still true concerning animal emotion.

But they had no doubt that living with Lyudmila had accentuated the domesticated behavior of Pushinka and her family. They had all learned to be quite a bit more dog-like. Lyudmila had also observed what she thought were signs that Pushinka was exhibiting a rudimentary form of reasoning ability.

A particularly memorable instance of this was a sly trick that Lyudmila witnessed Pushinka play on a crow—a trick that fooled Lyudmila as well. One day when Lyudmila was on her way back to the house from spending some time with the foxes at the farm, she saw Pushinka lying perfectly still in the grass in the back yard of the house. She looked like she wasn't breathing. Terrified, Lyudmila rushed over to her, but Pushinka remained totally still, and even with Lyudmila so close, showed no signs at all of breathing. Lyudmila turned to rush to get the vet. Just as she turned, she noticed a crow fly down onto the yard near Pushinka. In an instant, Pushinka sprang to life and grabbed the crow. How, Lyudmila thought to herself, could such clever planning be explained if Pushinka wasn't capable of some sort of simple rational thought? Her performance suggested that she understood the crow would see her as being dead, and seemed to involve a basic understanding also that some crows like to feed on dead animals. If so, her trap had been set brilliantly.

Perhaps the most astonishing instance of what seemed to be a kind of inference by one of the foxes happened when Marina, an assistant who had come to help with the work at the new house (not Marina, Lyudmila's daughter), sat down in the house to smoke a cigarette, as she did every day. One of the foxes in the house, which she had nicknamed Jacquelin, was especially taken by Marina and the feelings were mutual. When Marina sat down for her smoke that day, the ashtray that was usually on the table where she was sitting wasn't there. She asked the others in the house whether they knew where it was, and everyone started to search for it. Suddenly they heard some noise from behind a cupboard in the room, and out came Jacquelin pushing forward the lost ashtray. They were all amazed.

Perhaps it was sheer coincidence, and Jacquelin had simply stumbled on the ashtray and was playing with it as if it were a toy. But it certainly seemed that she had understood what Marina was looking for. Perhaps she had made the connection by having observed Marina smoking so many times. Lyudmila had no way of getting inside Jacquelin's mind, so she couldn't pursue this hunch. In the coming years, a researcher who did have expertise in animal cognition would learn about the foxes and travel to Akademgorodok to conduct a fascinating study that demonstrated just how strong their ability to make inferences from observing people is.

WHAT LYUDMILA AND DMITRI WERE EQUIPPED TO investigate further was the other ways that innate traits and learning might be affecting their tame foxes. They were constantly availing themselves of the latest techniques for research, and during the time Lyudmila was living at Pushinka's house, she and Dmitri decided to see whether they could delve even deeper into what degree the behaviors they were seeing in the tame foxes were genetically based.

Even as they tried to hold all conditions constant for the foxes, there were subtle, almost imperceptible differences that could creep into an experiment. For instance, what if the tamest mothers treated their pups differently than the aggressive moms treated their pups? Maybe pups *learned* something about how to be tame or aggressive toward humans from the way their moms treated them?

There was only one way to confirm for certain that the behavioral differences they were seeing *between* the tame and aggressive foxes were due to genetic differences. Dmitri and Lyudmila would have to try what is known as "cross-fostering." They'd have to take developing embryos from tame mothers and transplant them into the wombs of aggressive females. Then they would let the aggressive foster mothers give birth and raise those pups. If the pups turned out tame themselves, despite having aggressive foster moms, then

Lyudmila and Dmitri would know that tameness was fundamentally genetic and not learned. And, for completeness, they would also do the same experiment with the pups of aggressive mothers transplanted into tame mothers to see if they got parallel results.

In principle, cross-fostering was straightforward; researchers had used the procedure to examine the role of nature versus nurture for many years. But in practice it was easier said than done, it was technically difficult to pull off, and it had worked much better with some species than others. No one had ever tried to transplant fox embryos. Then again, no one had tried lots of things they had done, and so Lyudmila decided she would have to learn this delicate procedure on her own. She read all she could on transplant experiments that had been done in other species, and she conferred with the veterinarians they had on staff. Lives were at stake, so she took her time, learning everything she could.

She would be transplanting tiny, delicate embryos—on the order of eight days old—from the womb of one female into the womb of another pregnant female. Some of the embryos from tame mothers would be transplanted into the wombs of aggressive mothers, and some of those of aggressive mothers would be transplanted into the wombs of tame mothers. When the pups were born seven weeks later, she would closely observe their behavior to see if the pups of tame mothers became aggressive and if the pups of aggressive mothers became tame. But how in heaven's name was she going to know which pups in a litter were the genetic offspring of the mother and which pups were the ones she had transplanted? Without that information, the experiment was futile. She realized that the foxes had their own unique color coding system. Coat color is a genetic trait, so if she carefully selected the males and females so that the coat coloring of their offspring would be predictable, and the pups of the aggressive mothers would have different colors from those of the tame mothers, she'd be able to tell which pups were the genetic offspring of a female, and which had been transplanted.

Lyudmila led the transplant surgeries with her faithful assistant Tamara by her side. Each surgery involved two females, one tame and one aggressive, each about a week into pregnancy. After lightly anesthetizing the foxes, Lyudmila made a tiny surgical incision in each female's abdomen and located the uterus, with its right and left "horn," each of which had embryos implanted in it. She then removed the embryos from one uterine horn and left the embryos in the other. Then she repeated the procedure with the second female. She transplanted the embryos that had been removed from one mother into the other in a drop of nutritional liquid that was placed into the tip of a pipette. "The embryos," Lyudmila recalls with the pride of a job well done, "stayed outside the uterus [at room temperature from 64 to 68 degrees Fahrenheit] for no more than 5–6 minutes." The females were then moved to a postoperative room and given time to recover.

Everyone at the Institute anxiously awaited the results. Even with the surgeries having gone so well, the transplanted embryos might not survive. Their wait paid off. It was the caretakers who were the first to discover the births of the first litters, which was often the case with new developments with the foxes. They sent word right away to the Institute. "It was like a miracle," Lyudmila recorded. "All the workers gathered around the cages for a party with wine."

Lyudmila and Tamara began recording the pups' behavior as soon as they left their nests and began interacting with humans. One day Lyudmila watched as an aggressive female was parading around with her genetic and foster pups. "It was fascinating," Lyudmila recalls, ". . . the aggressive mother had both tame and aggressive offspring. Her foster tame offspring were barely walking but they were already rushing to the cage doors, if there was a human standing by, and wagging their tails." And Lyudmila wasn't the only one fascinated. The mother foxes were as well. "The aggressive mothers were punishing tame pups for such improper behavior," Lyudmila recalls. "They growled at them and grabbed their neck, throwing them back in the nest." The genetic offspring of the aggressive mothers did not show curiosity about people. They, like their mothers, disliked humans. "The aggressive pups on the other hand retained their dignity," Lyudmila remembers. "They growled aggressively, same as their mothers, and ran to their nests." This pattern was repeated over and over. Pups behaved like their genetic mothers, not their foster mothers. There was no longer any doubt—basic tameness and aggression towards humans were, in part, genetic traits.

The house experiment with Pushinka had shown that tame foxes had also learned some of their behavior. Living with humans had taught the foxes additional ways of behaving, some of which they shared with their domesticated dog cousins. Genes surely played an important role, but the tame foxes were not simple genetic automatons; they learned to identify individual people and became particularly bonded to them, and even defended them, owing to the process of living with them. That these learned behaviors were so dog-like provided the tantalizing suggestion that wolves in the process of transforming into dogs might also have learned these behaviors by living with people. Dmitri and Lyudmila had produced some of the best evidence that an animal's genetic lineage and the circumstances of its life *combined* in generating its behavior, and had done so in a highly innovative way.

BACK WHEN DMITRI HAD FIRST EXPLAINED HIS plan for domesticating foxes to Lyudmila, she had thought about the moving words of the fox character in Antoine de Saint-Exupéry's classic story, *The Little Prince*. "You are forever responsible for what you tame," the fox tells the prince. She felt that responsibility intensely, as did Dmitri and her assistants, and to some extent all those at the Institute. This, in part, was why they had hired some night watchmen to keep an eye on the farm and its precious inhabitants. With that sense of responsibility had also come love. Living in the house with Pushinka and her pups, Lyudmila and her assistants had truly come to love their companions, every bit as much as dog and cat owners love their pets. There was no point, Lyudmila knew in her heart, in denying that. The powerful love they were feeling was also important in illuminating how the bond between people and animals had become so strong.

Inevitably, that love also carries with it great sorrow and loss.

On the morning of October 28, 1977, as Lyudmila and Tamara approached the experimental house, they didn't see the foxes peering out the windows, and as they got near the front door, they didn't hear them yammering with excitement. This was very odd; the foxes *always* greeted them. They anxiously opened the door. No foxes came running to jump all over them. As they walked into the house, they realized it was empty. Then they noticed blood marks all over the room, on the floors and the walls. Horrified, they realized that some thugs had broken into the house in the night and killed the foxes for their furs.

Lyudmila and Tamara were in shock. After a few moments of stunned silence, they burst out crying. Then, suddenly, they heard some whimpering, and to their great joy, little Proshka, the most timid of Pushinka's grandsons, came scampering into the room. "When Proshka heard our voices," Lyudmila remembers, "he came out of his hiding and didn't leave our side." The quietest of all the foxes, the one who was such a loner, had been clever and lucky enough to survive.

Proshka required special attention for some time in order to return to his normal self, and then continued living happily in the experimental house. More foxes were brought to the house, and before long, they had pups, including one who was named Pushinka II. And the foxes continued to live in the house for several more years, with no further incidents. Lyudmila, though, spent less time there. She had other work and it was simply too painful.

How the murders could have happened remains a mystery to this day. The house was surrounded by a high fence, and the doors of the house, which were locked, had not been tampered with. The two night watchmen who patrolled the fox farm reported nothing unusual. The police were called in. They were tight-lipped about their investigation—this was the Soviet Union in 1977—but they talked with Lyudmila and Dmitri and they questioned the workers. No one thought they were involved, but they might have seen or heard something. They hadn't. The murderers, apparently, had come and gone quickly in the dead of night.

"Almost 40 years have passed," Lyudmila says today, "[but] I am still horrified." "One of the reasons for this tragedy," Lyudmila says, "is that our foxes trusted people, they didn't know that besides people who love and pet them, there are some people that can shoot them."

She was grateful that others took on increasing responsibility for continuing the house experiment. Finding it so painful to spend time at the house, she moved on to launching a new set of illuminating studies with her very special foxes.

6

Delicate Interactions

The cross-fostering genetics experiment combined with rapid development of the close bond between Lyudmila and Pushinka was like the evolution of the human-dog relationship being accelerated to warp speed. That the artificial selection for tameness could catalyze such a profound change in an animal's behavior, from the natural inclination to live as a loner in adulthood to forming such strong attachment, and with an animal from another species no less, is remarkable. How quickly this same change came about in wolves is impossible to know, but both genetic and archeological evidence suggests that a deeper bond than we developed with any other animal formed between us and wolves, or wolf-like proto-dogs, at least thousands of years ago, and perhaps tens of thousands of years ago. So close has the relationship been for so long that some experts argue that our two species coevolved, meaning that we acquired genetic adaptations to living with one another. Life with dogs, it seems, has been bred in our DNA, and life with humans has been bred into theirs.

A powerful testament to how far back in time the human-dog bond developed, and how strong it quickly became, is the wealth of ancient dog burials that have been discovered all over the world. Many of our prehistoric ancestors buried their dogs in graves just like the ones in which they buried their human loved ones, and sometimes in the same grave with their human masters. In fact, they began doing so right from the time dogs are generally thought to have been first fully domesticated, about 14,000 to 15,000 years ago.

The earliest find of a dog burial so far, dated to between 14,100 and 14,600 years ago, was discovered in the German town of Bonn-Oberkassel. The grave contained the fragmentary remains of a female dog buried along with the bones of a man aged fifty and a woman aged twenty, thought to have been the dog's masters. A burial that evokes the closeness of the relationship more vividly is a site in the Jordan Valley dating to 12,000 years ago. A grave was found at the entrance to a home, marked by a large stone slab, with a human skeleton curled up on its right side, ritually placed in a sleeping position, with the left arm stretched out and resting on the skeleton of a dog puppy, as if in an embrace. A number of dog burials that indicate the great importance of dogs in the life of one community were found at a site in Siberia, on the shores of Lake Baikal, dating to 7,000 to 8,000 years ago. There the dogs were clearly laid to rest with great care, and some were buried with valuable objects. A number of dogs were buried with spoons and knives carved from deer antlers, and one was buried with a necklace made of deer teeth around its neck, as worn by the people in the area. In one grave, a man was buried with two dogs, one laid out on each side of him.

These graves suggest that while dogs were undoubtedly of great use to these early societies, likely as pack animals, guards, and partners in hunting, the nature of the relationship had developed far beyond a purely utilitarian affair. Many experts believe these burials indicate that dogs were seen as spiritual beings who should be treated with the same respect in death as humans.[1] Good support for this comes from the Lake Baikal site, as not only were valuable items buried with the dogs there, but the people who lived there were foragers, relying heavily on fish and seals from the lake for their subsistence, so they probably didn't need their dogs to help them hunt.

Why were our ancestors so enamored of dogs, and why did they hold them in such esteem? One reason may be that for thousands of years, they were the only wild animal to become domesticated, which would have been reason to believe there was something special about them. Conservative estimates that dogs were domesticated 14,000 to 15,000 years ago make them *the* domesticate for about 5,000 years, until sheep and cats also made the transition sometime around 10,500 years ago, followed in comparatively rapid succession by goats, about 10,000 years ago, and both pigs and cows about 9,000 years ago.[2]

A number of recent archeological finds suggest that dogs and humans have lived together for many thousands of years longer than was previously thought, and some intriguing new findings in genetics suggest that in the course of our long time together, we became increasingly good for each other's well-being. Perhaps the most evocative of the archeological finds is a set of fossilized footprints on the floor of Chauvet cave in France, famed for its elaborate wall paintings of fierce predators, including lions, panthers, and bears, which date to approximately 26,000 years ago. Alongside a trail of prints left by a boy, estimated to have been about ten years old, run another set of prints, those of a large canid animal, and the prints suggest that this may have been an animal more like a dog than a wolf.[3] It's enchanting to imagine that boy with his proto-dog loyally trotting along by his side, and the images of fearsome predators that line the cave's walls leave little doubt about why a wolf-dog companion would have been welcomed. A still earlier date for the presence of dogs, or dog-like ancestors, in our lives has been suggested on the basis of a dog-like skull found at another cave, in Belgium, which dates to approximately 31,700 years ago.[4]

As we lived together for so many eons, through so many changes in our environment and lifestyle, with humans developing from hunter-gatherers, to farmers, to urban dwellers, and our dogs coming along with us on the journey, our genomes adapted in complex and similar ways, both to each other and to the environment. For example, genetic adaptations similar to those in the human genome that allowed our ancestors to begin eating starchy foods, like the wheat, barley, and rice they domesticated, also appear in the dog genome, and they allowed dogs to eat these foods as well, perhaps first having scavenged them from our ancestors' fields or stockpiles, and later being fed them. Wolves, who eat a meat-heavy diet, don't have the complex genetic machinery to eat these grains.[5]

That we adapted specifically to life with one another is also attested to by a number of positive effects we have on each other. Many studies have shown that living with dogs has many beneficial physical and psychological effects on us, such as lowering our blood pressure and rates of heart disease, as well as the frequency with which we go to doctors, and increasing our general sociability, while also helping us fight off depression. Recent work on the neurotransmitter oxytocin confirms what every dog owner already knows—that we and our dogs genuinely enjoy each other's company. Both sides feed off it in a positive feedback loop, in a kind of feel-good snowball of mutual reinforcement.

Researchers have known for more than four decades that oxytocin is fundamental in the bonding between human mothers and their children (as well as in nonhuman mother-child bonds).[6] More recent work has found that when a human mother and her newborn engage in mutual gazing, oxytocin levels in the mom rise, and the newborn's oxytocin system kicks into high gear. This leads to more gazing from the infant, which again increases the mother's oxytocin level.[7] When this work was published in 2014, we already knew something about the role that oxytocin plays in dog-owner interactions: when we pet our dogs, oxytocin levels rise in both us and them.[8] But now we know even more: a 2015 study has shown that the mother-child oxytocin loop that turns on as a result of human mutual gazes is also at play with owners and their dogs. This study found that when dogs and owners simply gaze at one another, oxytocin levels go up in both. This leads to increased petting and more oxytocin in response to that petting, in a chemical lovefest. What's more, if you spray oxytocin up a dog's nose, and researchers do, it gazes longer at its owner, setting off another lovefest. None of this happens when you replace dogs with wolves in this experiment, the discovery of which must have required steely bravery on the part of the researchers.[9]

These biological effects dogs and their humans have on one another are brought about by changes in the genes that control the production of hormones and neurochemicals in our systems. They constitute additional strong support for Dmitri Belyaev's theory that selection for tameness would unleash a cascade of changes in the production of the chemicals that regulate bodily functions. Dmitri had emphasized changes in the production of hormones in his theory at the start, because much less was understood about neurochemicals, like oxytocin, when he first formulated his theory. As research in the 1970s began revealing the powerful role they play in regulating an animal's behavior, particularly illuminating their effects on how happy or depressed an animal is, Dmitri realized that they might also be integral to the changes brought about by destabilizing selection. The rapidly emerging understanding of how sensitive animal behavior is to changes in the levels of these chemicals coursing through our brains and throughout our bodies helped to explain why the behavior of the tame foxes had changed so rapidly, and why Lyudmila and Pushinka had developed such a strong bond.

For the first decade of the fox experiment, Dmitri and Lyudmila had not been able to investigate much about how the biochemistry of the tame foxes was changing. Their finding of much lower levels of stress hormones in the tame foxes was a strong start. But more work had to wait for methods of measuring, and manipulating, the levels of these chemicals to be developed. As great progress was made on this front in the 1970s Lyudmila and Dmitri were able to make many more important discoveries.

ONE OF THE IMPORTANT NEW FINDINGS CONCERNED the neurochemical serotonin. Discovered in the 1930s, serotonin was first identified as a muscle constrictor, helping to tone muscle, hence its name, which is shorthand for "toning serum."[10] But in the early 1970s, the fact that higher levels of serotonin in the brain lead to an elevated mood and less anxiety was discovered, and in 1974, the year Lyudmila and Pushinka moved into the house, Prozac, the first serotonin-based antidepressant, burst onto the scene. The new understanding of serotonin's effects led Belyaev to believe that the tame foxes might appear so happy in part because they were producing higher levels of the neurochemical. Lyudmila ran tests of the serotonin levels in the blood of the control and tame foxes and found that, sure enough, the level was significantly higher in the tame foxes. Not only did they seem happier, they were, or at the very least hormones suggest they were. The same is true for dogs compared to wolves: the former have much higher levels of serotonin.[11]

Another obvious candidate for Lyudmila and Dmitri to look into in their foxes was the hormone melatonin, which was known to regulate the timing of mating and reproduction in many species. It must have been involved, they speculated, with the elite females going into estrous earlier, and with the few cases of those who had done so more than once a year. Melatonin was thought to be involved in the timing of mating in animals because, in the wild, many animals begin to mate when the days begin getting longer, and melatonin production changes according to the amount of light an animal is exposed to. It rises and falls both from day to night and seasonally. During the day, the level goes down and when night comes, it rises. The change in the levels in an animal's system, as days start getting longer again from winter to spring, was thought to be one trigger for mating in many species.

The control mechanism governing these changes in melatonin production is the tiny pineal gland, which is a light receptor buried deep inside the brain. For that reason, it was dubbed the "third eye," and it was thought to be vital to life functions because it is located near the very center of the brain. Back in the seventeenth century, René Descartes had even conjectured that it was the "seat of the soul," where thoughts are generated.[12] But precisely what the gland did, other than sensing light, had remained a mystery. At last scientists discovered that it manufactures melatonin, as well as a number of other hormones. Researchers also found that changes in levels of melatonin were involved in the production of the sex hormones critical to the process of mating and reproduction, kicking their production into high gear.

Dmitri and Lyudmila decided to investigate whether changes in the amount of light the foxes were exposed to would affect the timing of when they became ready for mating. During the fall months, Lyudmila and her assistants exposed a group of both elite and control foxes to two and a half more hours of light a day than was normal for that time of year. At the start, Lyudmila didn't have the technology for measuring their levels of melatonin, which was a very tricky procedure that had only recently been developed and required sophisticated expertise. But she did have the ability to measure the levels of sex hormones, which is much less complicated. She and her team ran an analysis and discovered that with increased light exposure, sex hormone levels did increase significantly, in both the control and elite foxes, but the effect was *much* more pronounced in the elites. What's more, this was true for both females and males, which was one of the first significant changes in the sexual biology of the males that Lyudmila had seen. In fact, she found that the levels were so high in some of the tame foxes that when she examined them she discovered some were ready for mating, and this time, that was true for some males as well as females. This was a big new first for the fox study, as now Lyudmila could examine whether these elites could *conceive* more than once a year, one of the most radical changes brought about by domestication in other species. She carefully selected pairs for mating, but none of the females became pregnant. Clearly more was involved in regulating the reproduction process than the elevation of sex hormones.

Still, this was a major finding. It suggested that the tame female foxes, who had already been going into estrous earlier without the special exposure to more light, were producing different levels of melatonin than the rest of the foxes in response to the same amount of light. Whether they were producing more or less couldn't be determined without measuring the levels of melatonin in the foxes' systems. This was a conundrum. They'd have to find someone with the expertise. While one of the researchers at the Institute, Larissa Kolesnikova, was a specialist in the workings of the pineal gland, even she didn't know the sophisticated methods for measuring melatonin.

Dmitri asked Larissa if she would be willing to join the fox team to work on this study and to get the training to do it. She'd have to travel outside of the Soviet Union for that, he told her, and the training would take several months. Larissa was intrigued by the challenge, allured by the opportunity to make a significant discovery. She also found the prospect of working closely with Dmitri Belyaev of great appeal, recalling, "you know, there was the attraction to working with him . . . more attractive than my feeling of insecurity."[13] She agreed to go. But sending her overseas wasn't so simple. Now Dmitri had to get her the clearance to travel and find the funds to pay for the training. Despite the isolation and the relative paucity of funding that the Cold War had imposed on Russian scientists, he was determined to be at the forefront of that work, and he had the power as the head of a major institute to get things done. He arranged for Larissa to travel to the University of San Antonio Health Center, where cutting edge work on measuring melatonin levels was underway.

Learning the technique was only half the battle of measuring the melatonin levels, though. Larissa would have to take samples of the foxes' blood, both during the day and late at night, right before the normal reproductive season in late January, when the key changes in melatonin production were thought to occur. Getting samples during the day wouldn't be any particular ordeal, but winter nights in Siberia are often brutally cold. The temperature regularly drops to −40°F. Larissa told herself she'd just have to focus on the beauty of the nights, with moonlight glinting off the snow, turning it, as she recalls, "bluish, lilac, and purplish shades," and a stunning display of stars "looking so far, far away."[14] But there was another problem. She couldn't do the work alone, the caretakers would have to help her. They had helped with this sort of work before, measuring stress hormone levels. But that was done only during the day.

Most of the caretakers were women with families to take care of. Larissa would be asking them to leave their homes and families and come to the farm for several hours, from 11 p.m. until 2 a.m., for two weeks. She fondly recalls, "I do not remember anybody complaining that she could not put a kid to the bed, or would not have time to cook for the next day . . . Their motto was 'If it is for science, let's do it.'"

On a bitterly cold night, the driver of the Institute van, a genial fellow named Valery, picked Larissa up at her apartment in Akademgorodok a few minutes before 11 p.m. and then headed over to the little town of Kainskaya Zaimka to pick up the workers. Larissa remembers that every one of them was waiting for the van, standing at the windows watching for it to pull up. They knew timing was tight and they didn't want to be responsible for slowing things up.

When he pulled up next to the rows of sheds at the fox farm, Valery put the van in park and kept the engine running, settling in for a nap while Larissa and the others considered the list Lyudmila had written that day of the foxes to be sampled. They had to map out their route for the night so that they could move as fast as possible. Heavy snow had just fallen, and the first thing they had to do was shovel paths to the sheds and over to the lab, where they would carry the foxes for the sampling. The night was virtually pitch black, with very little moonlight, so some of the women had to hold flashlights Lyudmila had provided. Making their way through the sheds, they hurried to locate the right foxes, shining the flashlights on the nameplates above their cages. Then holding the blessedly warm foxes tightly in their arms, they rushed from the sheds to the lab and back, as though engaged in some kind of clandestine military operation. When the sampling was done, they all ran to the van and Larissa recalls how, "Valery opened the door for us, laughing and asking if we were completely frozen or not yet."

After the blood samples were analyzed, Larissa met with Lyudmila and Dmitri and told them that she'd discovered a curious thing: the amount of melatonin circulating in the blood of the tame foxes was no different than in the control foxes. What was different, though, was the level of melatonin in the pineal gland. It was much higher in the tame animals.[15] This result was very strange, she said. The tame foxes were making much more melatonin, as expected, but it was amassing in their pineal glands in a kind of crystalized form so that it was "stuck," unable to make its way to the bloodstream. The elites' pineal glands were also found to be much smaller, about half the size, of those of the control foxes. Exactly what could be causing all this, none of them had any idea.

Clearly dramatic changes were happening to the endocrine system, the system responsible for the creation of hormones, in the tame foxes. But due to the limited understanding of the workings of that enormously complex system, it was impossible to say precisely what was happening and why. So complex is the endocrine system that it's still difficult to interpret this finding today. What could be said, because of the stark differences of the effects in the tame versus the control foxes, was that simply selecting foxes for their tameness had led to profound and complex changes to their reproductive system, just as Belyaev had conjectured so many years earlier.

WHILE DMITRI AND LYUDMILA WERE conducting their investigations into hormone and serotonin levels, Dmitri was also furiously planning for the fast-approaching International Genetics Congress, to be held in Moscow in August of 1978. As the Secretary General of the Congress, he was in charge of all of the arrangements, and he wanted it to be a gala affair, showcasing the finest offerings of Russian culture as well as the very best of the newest research from around the world, and from within the Soviet Union. The meeting would bring together researchers from sixty countries—3,462 geneticists in total—almost none of whom had ever been to the Soviet Union. This was a huge coming out party for Soviet genetics, a chance to show the whole world that they were well out from under Lysenko's boot and were doing first-class work. Dmitri wanted participants to never forget their trip to Moscow, and to come away with a vastly different impression of the Soviet Union than the one given by news broadcasts, which usually covered the latest conflict of the Cold War.

Détente had paved the way for this unprecedented opening of the doors to Western genetics, and in another sign of the seriousness of the Soviet authorities about new cooperation with the West, the USSR Academy of Science and the US National Academy of Sciences teamed up in 1977, the year before the International Genetics Congress, to conduct an assessment of the quality of Soviet research programs. They appointed John Scandalious, a senior geneticist from North Carolina State University, to visit a number of Soviet centers in genetics and evaluate them. The Institute of Cytology and Genetics in Novosibirsk was on his list, and his visit offered Dmitri a dry run at putting Russia's best foot forward.

Scandalious was put up at the lavish hotel in Akademgorodok reserved for visiting dignitaries and wined and dined, with many evenings of free-flowing caviar and vodka toasts. Dmitri and his wife, Svetlana, had him over to their home for several of their signature dinners with the Institute researchers, featuring much storytelling by Dmitri and lots of lively debate. Scandalious was impressed by how intensely the researchers at the Institute were thirsting for knowledge, not only about the latest scientific findings in the West, but about culture and politics as well.

Belyaev proudly brought him to the fox farm for a visit and Scandalious warmly recalls how Dmitri gently took one of the tame foxes from its cage and "handled it like it was a little baby, stroking it and talking to it." Scandalious had seen Dmitri as a somewhat stern man at first, and had then seen his warm side in spending more time with him. But even so, he was surprised to see him with the foxes, suddenly so gentle and affectionate. He was also struck over the course of his visit by how unofficious Dmitri could be when it was for the betterment of science, and how concerned he was for his researchers. One day as the two headed out of a meeting that had annoyed Dmitri, he said to Scandalious, "That guy is a pompous ass." "When we spoke science," Scandalious recalls, "Dmitri was very enthused, but at the same time depressed at how far back they were from the

West."[16] When Dmitri learned that a number of younger scientists at the Institute had given Scandalious unpublished manuscripts to submit on their behalf to European and American genetics journals, which was still a violation of official rules, all Dmitri said to Scandalious was that it was okay and he needn't worry about being searched on the way out of the country.

Belyaev and the Institute received high grades in Scandalious's report, and Dmitri felt confident that was a good omen of what the International Genetics Congress would bring.

Indicative of the stature he now held in Soviet science, Dmitri had secured approval for the Congress to be officially opened at the Kremlin, the heart of Soviet power and legend. Within the imposing towered walls lay the seat of the Senate, the Great Bell Tower, the Tsar's Cannon, the Arsenal, the Armory (treasury), and a number of exquisite churches with breathtaking gold turrets. The opening session of evening addresses was held in the cavernous main theatre of the State Kremlin Palace, which seats 6000.

The president of the International Congress of Genetics, seventy-nine-year-old botanist Nikolai Tsitsin, took the stage first and immediately assured the august assembly of the world's leading geneticists that the Soviet Union was back in the business of serious science, beginning his speech with the welcome, "On behalf of the Soviet people, scientists, geneticists and selectionists. . . ." His use of the word "selectionists" sent an unmistakable message that Lysenko and his denialism were dead and buried and that Gregor Mendel's genetics and Charles Darwin's theory of natural selection were again the driving force of Soviet genetics. Dmitri Belyaev could not have been more pleased. Making that case was one of his main objectives in signing on for the massive undertaking. The president also made a special point that night of noting that Darwinian ideas on natural selection had recently been enhanced by Professor Belyaev's powerful new theory of destabilizing selection.[17] Yes, Dmitri thought, the meeting is off to a good start.

After the opening speeches, the guests headed to a lavishly laid out banquet hall in the Kremlin State Palace, where as one of the attendees recalls, "champagne and black caviar were not limited."[18] On other nights, Dmitri and Svetlana hosted late night cocktail parties in their luxurious suite at the Rossiya Hotel, which was listed in the Guinness Book of World Records as the world's largest hotel. With 3,200 rooms, the finest of which overlooked the Kremlin, it was equipped with its own police station. John Scandalious made sure to attend one of these fêtes, and his wife Penelope, who'd come along for this exotic trip, fondly remembers the spirit of international camaraderie in the room, the plentiful caviar, sturgeon and top-quality cognac, which was served with sugared lemon slices as chasers, and the many toasts to friendship and genetics.

As the Secretary General of the Congress, Dmitri was slated to give one of the evening keynote addresses, and naturally chose to discuss the fox experiment. After introducing all of the latest findings, he presented a short film that showed the foxes in action. He had hired a professional film crew to come to the farm, and Lyudmila and her assistants had brought them all around, showing them the tame foxes and how they responded so gleefully to attention, and also the aggressive foxes and how fierce they were. Lyudmila also brought the crew over to Pushinka's house to meet the current brood living there and how they came trotting right into the house from the yard when they were called.

The lights went down and as stock footage of cows grazing, horses prancing, and puppies frolicking around in a field rolled, a narrator announced in crisp English, "Domesticated animals have been bred by man for about fifteen thousand years." Then a little charcoal-colored fox appeared on the screen, prancing happily along a country road, off leash, accompanied by a woman in a white lab coat—one of the Institute researchers. The fox was sniffing the grass by the side of the road, wagging its curly tail, and glancing up at the woman repeatedly to make sure she was keeping pace—looking exactly like a dog. As the camera toured the farm, fox pups nibbled playfully at a researcher's fingers, adults wagged their tails excitedly as Lyudmila and her assistant Tamara walked by their cages saying hello, and the family of foxes living in Pushinka's house followed

Lyudmila out the door and into the back yard, gathering around her and competing for her attention. As the lights came back up, the room was abuzz with whispered comments about these amazing animals.

Dmitri completed his talk by informing the crowd that after the twenty years the experiment had been running, the farm now housed 500 domesticated adult female foxes, 150 adult males, and 2,000 young foxes, many of whom displayed the domestication traits. He closed by offering a tantalizing thought: that his theory of destabilizing selection and domestication "can also, of course, apply to human beings." He said nothing more on the subject, and discussion as the crowd streamed out of the auditorium was rife with speculation about what he was suggesting.

THE IDEA THAT HUMAN EVOLUTION MIGHT have followed essentially the same course as the domestication of dogs, goats, sheep, cows, and pigs was provocative to say the least. Were we humans really, in essence, domesticated apes? Some astonishing genetic analysis of humans published only a few years before the Moscow congress had proved that we were extremely closely related to the primates thought to be our closest relatives, chimpanzees. In fact, this work suggested that we were so closely related that genes alone could not explain all of the substantial differences in our physiology, not to mention our cognitive abilities.

In 1975, Mary-Claire King and A. C. Wilson published a *Science* paper where they note that "the sequences of human and chimpanzee polypeptides examined to date are, on the average, more than 99 percent identical." They hypothesized that this meant that the differences between the two species must be due primarily to changes in the regulation of the activity of genes, rather than to a series of new mutations that selection had acted on.[19] This argument fit nicely with Dmitri's theory of destabilizing selection. Dmitri had asserted that the dramatic changes involved in domestication were not due primarily to an accumulation of new genetic mutations favored by selection—though he knew that these surely played some role—but rather to alterations in the expression of existing genes that caused them to produce different results. Belyaev's core insight that the activity of genes could be turned on and off, or altered somehow, so that the same genes produced different results, such as tame behavior, curly tails, and the emergence of new coat colors, was being confirmed.

The term *gene expression* had begun gaining currency as researchers discovered more about just what a complex process the translation of the coding of a gene into a biological product, like hormones, is. As sequencing techniques improved and the elaborate workings of cells became better understood, researchers began to discover that gene expression was not a matter of a lock-step, computer-like "reading" of the genetic code by the cell. The code could be modified and production could be either stopped or increased. Cell biologists had determined that the manufacturing of the proteins, hormones, enzymes, and the other chemicals that genes code for, conducted by the little structure in the cell called the ribosome, could be interfered with to produce more or less of any given chemical product. The expression of a gene came to be understood as, essentially, the process by which the gene led to the production of more or less of a protein, hormone, enzyme, or other chemical by cells. And small changes in expression could produce large effects in an animal's physiology and life functioning. Some change or set of changes in gene expression, Dmitri thought, might explain why the pineal glands of the tame foxes were now producing so much more melatonin, and also why, even though that melatonin wasn't making it out into their blood systems, it seemed to be dramatically affecting the foxes' reproductive behavior.

Subsequent research has revealed that the expression of genes can be interfered with in a host of ways, and by multiple culprits, including environmental factors. The action of light in regulating the production of melatonin is only one of a multitude of examples.

The timing of when genes become activated can also be modified. For example, small bits of "noncoding DNA" that produce no product of their own can tinker with gene expression, causing certain genes to become activated either earlier or later in the development process. Some such change in the timing of activation probably explains one of the physical changes that had begun appearing in more and more of the foxes during the course of the 1970s. The white star that had appeared on the head of only one male pup in 1969 started showing up on the heads of more foxes with each new generation. Advances in embryology allowed researchers in that field who worked at the Institute of Cytology and Genetics to explain how the stars were emerging. By analyzing the hair that constituted the stars closely, Dmitri and Lyudmila had found that the stars were made up of only three to five white hairs. Their ongoing pedigree analysis had shown that the pattern of inheritance by which the stars were appearing indicated they were not due to new genetic mutations; the number of stars was increasing much too rapidly for that to be the case. Something else was going on, and the embryologists figured out that it had to do with a change in timing in one aspect of the development of the fox embryos.

By this time, embryologists had worked out ways to track the migration of cells to different parts of the body as an embryo develops. Some migrate to the top of the spinal column and become brain cells, some migrate to become lung cells, some heart cells, and so on. The embryologists at the Institute were able to determine that the white coloring of the few hairs creating the stars was due to the timing of when the cells responsible for the coloration of hair were being instructed to migrate and become skin cells. While these cells normally migrate during the period from 28 to 31 days of development, in the foxes with the stars on their forehead, the movement was delayed by two days. This delay caused an error in their production of hair color, leading the hairs of these cells to be white.

The timing of the cell migration must be governed, Dmitri and Lyudmila concluded, by a chemical produced by a gene, and it seemed that the expression of that gene had been affected by the destabilization brought about by the selection for tameness. One small example of what a delicate operation the functioning of genes is.

Much subsequent research has shown that gene expression is an extraordinarily complex affair. So complex and unpredictable are the processes by which gene expression is regulated that learning how to take charge of the process, in order to fight diseases and harness the body's healing powers, is a quest we'll be on for many years to come.

Dmitri and Lyudmila were to experience a sad truth of the mystifying complexities of these elaborate operations when they decided to try again to mate some of the tame foxes before the normal January mating period. Lyudmila had discovered that some of the elite males in addition to females were sexually active and prepared for mating in the fall as well as in the winter mating window. This was without any manipulation of their exposure to light. The change had emerged, as with the females, from the continuing selection for tameness. In the fall of that year, she and Dmitri decided to see whether some of these foxes would mate if they were put together, and whether the tame females would now become pregnant. A number of them did, and though some of them miscarried, a few of them gave birth successfully. This was another huge step for the experiment. The question whether or not, if foxes were bred to be domesticated, they would be able to mate more than once a year, as almost all domesticated animals do, was answered.

Everyone, especially Dmitri Belyaev, was thrilled. Lyudmila recalls that "when we produced the pups, Belyaev went to the Institute and arranged an emergency meeting in the conference hall." Dmitri enthusiastically told the staff, "Here are results you should be proud of. Here are results you can boast about."

But the sad truth was that being able to give birth outside of the normal cycle did not also include being able to nurture the newborn pups. The mothers were not producing enough milk to sustain their pups, and the little they did produce, they were reluctant to dispense. For the most part, they ignored their pups. Lyudmila and her team did all they could to care for the helpless little creatures, feeding them on a rigorous schedule with a dropper. But that was not enough. None of them survived.

As Dmitri had hypothesized so many years earlier, destabilizing selection had substantially altered the genetic systems of the foxes, but some components of the delicate cycle of reproductive readiness were now out of sync with one another. With dogs and cats and cows and pigs, over the longer sweep of evolutionary time in which their domestication unfolded, changes in the selection pressures they were living under, due to living in close proximity to humans, had recalibrated their reproductive systems so that mothers could produce more milk and developed nurturing impulses more than once per year. It made perfect sense that the new selection conditions would have produced this new calibration; giving birth more than once a year would be selected for by natural selection once more offspring could be fed and protected. And the breeders of the animals would also have selected for the ability. In the foxes, selection for tameness had progressed to the stage where the animals could reproduce more than once in a year, but not yet care for their young. In principle, the ability to produce milk and to be a good mother would be the next step. But, as Lyudmila says, the reproductive system "cannot just be changed overnight."

THE EARLY 1980S WAS AN EXTREMELY PRODUCTIVE PERIOD for the fox experiment in beginning to solve the mysteries of the deep biological changes going on with the foxes, but the decade would evolve into a very challenging time for the experiment.

With the Soviet invasion of Afghanistan in 1979, renewed tension developed between the Soviet Union and the Western alliance of nations, reversing the thawing of détente. President Jimmy Carter sent covert support to the Afghan rebels fighting the Soviet military, which was stepped up after the election of President Ronald Reagan in 1981, who also put new emphasis on building up US military might. The administration implemented the Reagan Doctrine, which involved the support of other resistance movements against Soviet influence in Latin America, Africa, and Asia as well as political and economic measures to undermine Soviet power.

The new tensions threatened to reverse the gains made in exchanges between scientists in the West and their counterparts behind the Iron Curtain, which Dmitri Belyaev had been so influential in furthering. Aubrey Manning, who had reached out to Dmitri and invited him to attend the 1971 ethology conference in Scotland, was distressed that barriers were being thrown up again between the scientific communities. "I just felt, this is preposterous," Manning recalls. "At that time the Cold War was absolutely at its pit," he says. "There was virtually no contact whatsoever between Russian scientists and Western scientists."[20] He decided that he had to make a statement of some kind, and he wrote to Dmitri that if he was open to a visit, he would like to come to Akademgorodok to see the foxes. They had stayed in touch, and he knew that many exciting developments had transpired since Dmitri had presented the experiment's findings in Aubrey's Scotland back in 1971.

Dmitri wrote back straight away to say Manning was always welcome—what's more, the Institute would pay all his expenses for the trip once he was in the USSR, which left Manning just the airfare to cover. "I wrote to the Royal Society [of London]," Manning recalls, "and said I thought it would be valuable to make contact. And they came up with a travel grant for me."

Manning arrived in the spring of 1983, and he recalls with a smile, "I was treated like royalty. They had so few visitors from the West at this time, that it really was quite something." Dmitri arranged a number of formal dinners and invited the heads of various of the Institutes of Akademgorodok and their spouses. These were sumptuous affairs featuring, as Manning recalls, "gigantic trays of delectable things." Not familiar with the Russian tradition of lavish, multicourse meals, he recounts how, "somewhat to my embarrassment" at the first dinner, after having eaten his fill he discovered the main course was still to arrive. He was also amazed that between the courses, people lit up cigarettes. He told Dmitri, "You know, this could never happen in Britain, because nobody is allowed to smoke until there's been a toast to the Queen, and that never happens until coffee is being served at the end of the meal." Dmitri immediately proclaimed, "I think it is time to toast the Queen, Aubrey!" Manning did the only thing he could and raised a glass "to the Queen!" At that point, Manning realized just how special a friend he had made. "I thought," Aubrey recalls, "it was a rather charming incident and very typical of Dmitri to make the whole thing into a joke. I loved it."

Aubrey was impressed by the science he was introduced to at the Institute of Cytology and Genetics. The researchers were top notch, and they knew so much more about science in the West than Western scientists knew of what was happening in the USSR. But it was more than just their knowledge of science that he was impressed by. So many of the people he met seemed well acquainted with Western culture. "One day, we were eating sandwiches on this boat," Aubrey recalls, "and I said jokingly, 'oh, how very English that is!'" On the boat with Aubrey was Dmitri's press secretary, Victor Kolpakov, who acted as an interpreter. "Without a pause," Aubrey recalls, "Victor said to me, 'there were no sandwiches in the market this morning, sir. Not even for ready money.'" Manning was floored. "That is a quotation from Oscar Wilde's *The Importance of Being Ernest*," he explains. He discovered that many of those he met were similarly acquainted with the major works of Western literature and could readily quote writers like Graham Greene, Saul Bellow, or Jane Austen. "It was quite extraordinary and very humbling really," he recalls.

This made the large gap in understanding between the West and the Soviets all the more tragic to him. Manning recalls a discussion of those tensions with Dmitri after dinner one night, when Dmitri invited him to the sanctum sanctorum of his home study. "He was smoking and we sat there chatting a bit," Aubrey tells it, ". . . about the fact that there still was a great deal of mutual distrust between the West and the Soviet Union." Dmitri responded, "Why is there this difficulty?" and Manning explained that the West felt dangerously threatened by the Soviet block. Dmitri was puzzled at this, asking "Afraid? Why are you afraid; there's no possibility of an attack," telling Manning that the Soviet Union was "a peace-loving nation." Aubrey recalls thinking of a quote he himself had memorized, of the Scottish poet Robert Burns: "Oh would some power, the gift to give us, to see ourselves as others see us, it would from many a blunder free us from foolish notion." On another occasion when the two talked politics, as Dmitri escorted Manning to a banya—a sauna where men sat around naked and shot the breeze—Dmitri turned to his friend and told him, "Aubrey, I think it a good thing if Andropov and Reagan take a banya together." Manning replied, "You're dead right," and he remembers thinking, "he's right, we are all naked human beings, the rest is less important."

The highlight of Manning's trip was his visit to the fox farm on a hot August morning. The foxes did not disappoint. "I remember this particular fox," Manning recalls, "was running around wagging his tail and coming up to me. And I was feeding it by hand and it was wagging his tail. It was astonishing." He played with a number of foxes and he was struck that "they did feel like dogs . . . they were like a slightly foxy dog . . . a bit like a collie would be."[21] While the foxes on the farm were delighted by his attention, Pushinka's descendants at the experimental house were another story. Galena Kiselev, one of the fox team members who was at the house when Dmitri and Lyudmila brought Manning over, remembers the visit well. The foxes weren't paying any attention to Manning, they were "gathered around me," she recalls, "trying to climb up my ankles, and looking into my eyes." Dmitri said to her, "Hey, Galena, what are you doing? Let them go to Manning." But, there was nothing that Galena could do. "The foxes that lived in Pushinka's house," she says, "they hated men and they loved women, because their caretakers were women." What struck Manning was not that they didn't care for him, but that they wanted to be loved by *any* humans.

At the end of the visit to the house, Dmitri and Lyudmila brought Manning to the most special place on the farm, the bench at the side of the house where nine years earlier, Lyudmila had been sitting with Pushinka at her feet when Pushinka had charged off into the night to defend her. Sitting on the bench with Manning, Lyudmila, Dmitri, and Galena shared many a fox story.

A few days after the visit to the fox farm, Aubrey headed back to Edinburgh. He was surprised when Dmitri showed up to accompany him to the airport. Directors of Institutes in the USSR simply didn't perform such lowly functions as seeing visitors off. Aubrey had come to understand that. But Dmitri would not entertain the idea of letting Aubrey leave without a final farewell directly from him. "There was a woman at the [airport] gate," Aubrey remembers, ". . . to see that we had boarding passes so we could go forward." Dmitri, of course, didn't have a boarding pass, and she told him that was as far as he could go. "He just quite gently but firmly pushed her to one side," Aubrey continues, and "continued walking onto the tarmac with me." Then, Aubrey says, Dmitri "hugged me and gave me a Russian kiss." He was surprised, to say the least. "You know," Aubrey says today, "I'd never been kissed by a man before. . . . I was profoundly touched by this . . . I had tears in my eyes."

The warmth of the reception he'd been treated to in the Soviet Union made his reception back home all the more dispiriting. Arriving back in Scotland, Manning was interrogated by the MI5, Britain's spy agency, about his visit. "I found it quite horrible," Aubrey says, and he politely told them to go to hell, that he was there as a scientist, and he wasn't going to answer any of the silly questions they were asking him about the "killer wheat" they thought the Soviets were developing.

A great deal more turmoil in the relations between the Soviet Union and the West would unfold before scientists on opposing sides of the Iron Curtain could resume a free exchange of ideas and develop the sort of mutual appreciation that Aubrey Manning and Dmitri Belyaev enjoyed.

7

The Word and Its Meaning

By the mid-1980s, more of the tamest foxes were displaying the distinctive dog-like behaviors Pushinka had first exhibited. They responded to their names and came to the front of their pen when called. Control foxes never did. To see what would happen when they were given a bit more free range on the farm, a very few were allowed for walks on a leash and they were well behaved, and even fewer were trusted to be allowed out of their cages off-leash, as Pushinka had been, because they would follow the caretakers around. Lyudmila recalls of one worker how "you'd never see her walking by herself, there was always a little fox following her."

Some of the foxes now also looked so much like dogs that Lyudmila was confident that their anatomy was changing in the same ways that wolves' anatomy transformed as they became dogs. In particular, the snouts of the tamest foxes looked shorter and more rounded, making their visages friendlier to go along with their friendlier behavior. They had begun to look so dog-like, in fact, that one of the elite foxes, a female named Coco, who was something of a favorite at the farm, was mistaken one day as a stray dog by a young man from a suburb of Novosibirsk in the vicinity of the farm. Coco then went on quite an odyssey.

Coco was so appealing in part because from early on, she made a lovable chattering noise that sounded something like "co co co co co." Lyudmila says fondly of Coco, "She gave herself her own nickname." Everyone on the farm had followed her fate with great concern her first few weeks after birth. She was so tiny and weak that it looked like she wasn't going to survive. Even after the veterinarian gave her glucose supplements and vitamins every day, and hand fed her milk, she was still failing. Every morning when workers showed up at the farm, their first question was "How is Coco?" Even staff members over at the Institute wanted daily updates.

One staff member, Galya, always told her animal-loving husband, Venya, who was a computer technician in Akademgorodok, about how Coco was progressing when she got home at night. The two of them had discussed that if the veterinarian determined Coco had no hope, they wanted to make their small apartment into a fox hospice, allowing her to die with loving humans caring for her. Lyudmila agreed that they could take her in, and when the word came from the vet that there was nothing more he could do, they came to the farm to collect her. To their great surprise, when they got her home, Coco perked up and began to eat more. Within days, she was a new fox, and miraculously, she survived. Rather than bringing her back to the farm, Lyudmila was happy for her to live with Galya and Venya, who had become deeply attached to her. Coco, in turn, would become deeply bonded to them, and especially to Venya.

Venya was so enamored of Coco that he wanted to bring her into work with him, but that wasn't possible. Every evening when he got home, he would take her for a long walk in the nearby woods, keeping a firm grip on her leash. Coco was fine with the leash, and behaved well. But one evening when Venya got stuck late at work and Galya was walking Coco, the fox spotted a man walking way off in the woods and bolted toward him, breaking free from Galya. In a moment, Galya lost sight of her. Coco probably thought that the male figure in the distance was Venya and shied away when she discovered otherwise. Galya called out to her, but Coco didn't return, and Galya rushed home, hoping to find Venya so they could search for her.

For the next several days, Venya went back to the woods frantically searching for his dear friend, asking anyone he encountered if they had seen Coco. Finally someone told him they had heard that a young man from the town had found a fox that looked like a dog and taken it in. But by the time Venya tracked him down, Coco was gone. Later they learned that the very first night, Coco had screamed and scratched at the man's door so relentlessly that he finally just let her out.

Venya then heard rumors circulating among the kids at a local playground that Coco had been picked up by a woman who lived in the same building as the young man who had first taken her in. Venya managed to get her name and went to her apartment, but she refused to open the door. When he pleaded with her that Coco was a special fox, part of an experiment at one of the institutes at Akademgorodok, she only opened her chained door a crack and said tersely, "I do not have it." But later that night, she apparently got nervous about holding on to such a special fox, and she also let Coco out. The odyssey still wasn't over.

Venya now got word that the kids at the playground had seen Coco with a local teenage boy, who was known to be a bully, but they said they didn't know the boy's name or where he lived. All they would say is that they thought he was about twelve years old. So with Lyudmila's help, Venya set up an appointment with the principal of the middle school, and he and Lyudmila explained the situation. Right then, the teachers were instructed to make an announcement to every class that Coco was a special fox and if anyone had any information that would help find her, they should say so. It paid off. The boy's name was quickly coughed up, and Venya and Lyudmila rushed to his apartment. They arrived just in time to find the boy's mother in the process of sedating Coco, apparently preparing to kill her for her beautiful fur. Venya tore Coco away from the woman and ran out to the street with her limp body in his arms. As Coco breathed the fresh air, she began to revive.

Coco lived happily in Venya and Galya's apartment for six more months, but when mating season came around, she became restless. She began scratching at the apartment door and keeping Venya and Galya up all night. Clearly she was longing to find a mate, so they consulted with Lyudmila and a worked out a plan. They would bring her back to the farm to mate, and then she would be moved into Pushinka's house. To smooth over the transition to a different home, she was first placed in the human half of the house, and then in time, joined the other foxes over on their side.

For years Coco lived in Pushinka's house, and Venya would visit every weekend, occasionally spending the night on a couch there. They also took regular walks together. When years later, Coco's health started to fail, Venya and Galya brought her back to their apartment to spend the last days of her life in their loving care. Lyudmila remembers Coco "behaving peacefully and spending that last period of her life very content and happy." Coco's greatest joy was sitting on a chair with Venya and looking out the window. On one such occasion she jumped off of the chair and fractured her right front paw. Shortly after that, she developed a bone sarcoma. Venya cared for her, but he knew it was the beginning of the end. Soon thereafter, Coco had a heart attack and died, with Venya and Galya by her side. They buried her, in a tradition we now know goes far back in our ancestry, on a small hill in the woods where she and Venya loved to walk.

Venya visits her grave regularly to this day.

THE SPEED WITH WHICH DMITRI AND LYUDMILA'S FOXES HAD BECOME an animal capable of being such a lovable pet was particularly striking given the natural inclination of foxes to live as loners once they mature to adulthood. This difference between foxes and wolves, who are such social animals, might be a key factor in why wolves were domesticated so much earlier than any other animal. We speak of lone wolves, but wild foxes are the loner of the two. The gap of thousands of years between the domestication of dogs and that of the several other animals—cats, sheep, pigs, cattle, goats, and more—suggests that something about the wolf ancestors of dogs made them especially well-suited for adjustment to life with human groups, and one idea is that the special factor was how social an animal wolves are.

The first wolves that laid by our ancestors' fires and shared their food not only were tamer than other wolves, they already had highly evolved social skills. Grey wolves live in strictly structured packs, typically containing 7–10 members (though packs can range up to 20–30 individuals), including an alpha—dominant—male and female. The family unit is central to the pack, which defends a large territory, using complex vocalizations to communicate with one another, as well with those in nearby packs. Bonds between group members are very tight, as can be seen during cooperative hunting and when pups nurse not just from their mother, but from other females in a pack.[1] Jane Goodall has argued that "[wolves] survive as a result of teamwork . . . They hunt together, den together, raise pups together . . . This ancient social order has been helpful in the domestication of the dog. If you watch wolves within a pack, nuzzling each other, wagging their tails in greeting, licking and protecting the pups, you see all the characteristics we love in dogs, including loyalty."[2] Their experience in cooperating with one another apparently equipped them for cooperating with us as well.

Dmitri thought that exceptional pro-social skills might have played a key role in the domestication of another species—*Homo sapiens*. While many animals, such as prairie dogs, parrots, and the leaf-cutter ants whose social lives E. O. Wilson described so memorably in *The Insect Societies*, live in close-knit social groups and watch out for one another's interests, we humans have distinguished ourselves as one of the most social species on the planet, especially if you include norms, cultural rituals, and forms of communication in the definition of sociality. The increasing strength of social skills and depth of social bonds were central features of human evolution from primate ancestors, facilitating our transition first to life in small family-based groups of hunter-gatherers and then to life in the more complex social environment of steadily larger and more complex inter-family communities. Dmitri thought that his theory of destabilizing selection provided a compelling explanation of what had set this transformation in motion.

The confirmation by the mid-1980s of many of Dmitri's conjectures about how domestication would follow from selecting for tameness in foxes emboldened him to take his thinking a big step further. He thought the time was now right for him to let the world know of his nascent ideas on destabilizing selection and domestication as they applied to human evolution. He had hinted at the end of his talk at the International Congress in 1978 that he thought his theory of destabilizing selection might offer insights about how humans had evolved from apes. Now he decided to develop his argument on this topic as the subject of the keynote address he would make at the next International Genetics Congress, to be held in India in 1983.

As a stream of groundbreaking findings about the course of human evolution made big news in the 1960s and '70s, Dmitri had formulated a theory about how humans had become such sociable beings; his work with the foxes suggested to him that we had, in essence, domesticated ourselves, and it all began with selection for tameness. His theory was based primarily on conjecture. But then, peering into the social lives of our ancestors in the prehistoric era, before they began to speak to us through the stories they carved in stone, is by necessity driven, at least at the start, by conjecture.

We may never know exactly when they first began to talk to one another, or to think in self-reflective ways, considered one of the hallmarks of distinctively human consciousness. We will never know for sure what stories they told one another around the fire at night, or exactly which songs they may have sung. But we do know that many forms of social ritual bonded them. They invested considerable care and time into the creation of works of art, making jewelry, carved figurines, and evocative paintings, such as images of distinctively human hands traced by red ochre paint, found on the walls of caves in many places around the world. Some of the oldest of these found to date, in the El Castillo cave in Northern Spain, date to approximately 40,000 years ago. Our ancestors also spent considerable time fashioning musical instruments, such as flutes carved out of animal bone, the oldest of which also date back about 40,000 years. And they saw their loved ones off in death by burying them with important items used in daily life, such as stone and animal bone tools, as the people living on the shores of Lake Baikal 8,000 years ago did with their dogs.

As Dmitri prepared to unveil his idea, the fact that a number of other proto-human species had evolved, and that some of them may even have lived alongside *Homo sapiens,* was only beginning to become accepted. The first Neanderthal fossils had been discovered way back in the early 1800s, but suddenly a number of major finds were painting a more complex picture of the proto-human family. Dmitri had read avidly about these discoveries and he thought his theory that we *Homo sapiens* had domesticated ourselves, fostering the strength of our social bonds, might explain why we were the one proto-human line that survived.

Louis and Mary Leakey made some of the most important of the discoveries of new proto-human species. Working in the Olduvai Gorge, in Tanzania, they uncovered a number of bones and skulls, as well as tools, which shed much light on the surprising diversity of proto-human lines. Mary made their first big discovery, in 1959, of the skull of a species that clearly evolved from a large primate.[3] But the shape of the skull was so different from that of the human skull that they concluded it could not be a direct progenitor of our line. It had an enormous jawbone and a pointed crest of bone running front to back along the top of the skull, known as the sagittal crest. Earlier researchers who had discovered skulls in South Africa with the same feature in the 1920s had concluded that this crest indicated that it acted as a kind of anchor for very strong, large muscles that ran down to the jawbone and were attached to it. Its bite would have been extremely powerful, which is why the Leakeys nicknamed the creature Nutcracker Man. The official name they gave to the species was *Zinjanthropus boisei,* meaning East African Man, and the discovery not only made front-page news all over the world, catapulting the Leakeys to fame, it caused a huge stir among experts in human evolution.

At this time, the idea that humans had descended from an ancestor in Africa was not widely accepted. Darwin and his colleague Thomas Henry Huxley had conjectured that our ancestors had likely evolved there because that was the only continent on which our nearest contemporary primate relatives are found. But paleontologists had discovered fossils of Neanderthals, in 1829, in Belgium, and then in other areas of Europe. The species got its name from the Neander Valley in Germany (*thal,* in German), where a Neanderthal skull was found in 1856. The skull of a different species that seemed to be proto-human was found in Indonesia, then called Java, in 1891, and given the moniker Java Man.

Excavations that began in the 1920s in a cave in China, near Beijing, produced another skull of this species, dubbed Peking Man because the city was at that time called Peking. This species was thought to have walked upright, so was named *Homo erectus.* Large piles of animal bones were found at the site, and some of them had char marks that suggest they resulted from cooking. Because the remains of different apparently proto-human species were found in such disparate locations, some researchers concluded that humans had evolved in various locations.[4]

In the 1960s, the Leakeys made their next major find, a jawbone and other fragments of a skull that was more human-like, as well as some hand bones. They argued the skull pieces found at Olduvai indicated that this species had a very large brain, and that the hand bones revealed it also had a good grip. They theorized that this species must have made some of the stone tools they had found in the area, leading them to name the species *Homo habilis,* "habilis" being the Latin for handle, nicknamed "handy man." Leakey and other researchers asserted that this species and Nutcracker Man, *Zinjanthropus boisei,* had coexisted, refuting the notion of a single line of evolution. Other anthropologists objected vigorously to this assertion, but as more fossils of both species were discovered, the Leakeys were proved correct.

One of the big open questions about the transition from more ape-like to more human-like species was when our ancestors had first started to walk upright. The Leakeys made important finds on that front, the first of which was a fossil of a *Homo habilis* foot bone, which suggested that it had walked upright. But the most startling evidence was uncovered at a site in the vicinity of Olduvai that Mary had begun to excavate a few years after Louis passed away in 1972. At this site, called Laetoli, Mary had in 1976 discovered a path of fossilized animal footprints, preserved in stunning definition by volcanic ash. When a colleague, Paul Abell, was investigating the prints one day, he noticed that one of them looked remarkably like that of a human foot. Further excavation revealed about seventy of these prints, looking eerily like a path of human footprints left in sand.

Perhaps no paleontological find has ever more evocatively transported us back in time. Close analysis revealed that the prints were made by three different beings, and that the toes, heels, and arches of their feet were in fact very much like those of human feet. There was little question that this species had walked upright, and the prints were made about 3.6 million years ago.

Because no fossilized bones of a hominid were found with the footprints, the species that had made them couldn't be identified. But evidence was that it must have been the species we now know as *Australopithecus afarensis*, the most famous member of which is named Lucy. A few years before the Laetoli footprints were discovered, the paleoanthropologist Donald Johanson spotted what was clearly an elbow bone jutting out of the ground at the excavation site he was investigating near the village of Hadar in Ethiopia. He and his team eventually uncovered the skull and much of the rest of a fossilized skeleton of a female hominid, which they named Lucy because during a party celebrating the find that night, a stereo nearby was repeatedly playing the Beatles song "Lucy in the Sky with Diamonds."

Just under four feet tall, Lucy likely had quite a small brain, given the size of her skull, but her skeleton also clearly indicated that she had walked upright. This was stunning for two reasons. One was how old the remains were, roughly 3.6 million years,[5] which was much earlier than paleontologists had thought our ancestors had first stood erect. The other was that anthropologists had expected that walking upright had evolved after the brains of hominids had grown more substantially. Lucy, some paleontologists now said, might have spent some of her time swinging through tree branches, as indicated by the size and shape of her shoulder bones. She was a remarkable link between the more ape-like species of proto-humans that had been discovered and our closer ancestors in the *homo* family. Lucy's bones dated to the same timeframe as the Laetoli footprints, and when Donald Johanson and his team compared the size and shape of Lucy's feet to those prints, some of them matched very closely.

Study of Lucy's bones and other bones of her species has revealed that *Australopithecus afarensis* children matured a good deal faster than human children do. So our evolution into humanness likely involved a delay of maturation, as was seen with so many of the characteristics of the tame foxes.

Belyaev thought the evidence suggested that humans had evolved largely through the process of destabilizing selection. In 1981, he published a scientific paper presenting this theory, and for his keynote speech at the XV International Genetics Congress in 1984—an honor given to the organizer of the prior Congress—he made the case in greater detail.[6]

In Belyaev's view, our ancestors had come under new stresses as their bodies and brains evolved. As they became more social animals, living in larger groupings required a moment-to-moment negotiation of a panoply of social interactions. The pace and complexity of the changes were not brought about primarily by the small increments indicative of natural selection operating on changes that resulted from single gene mutations. That had certainly played a role. But, he thought, the transformation through that process would have taken longer than the approximately four million years from the emergence of the earliest hominids, the Australopithecines, to modern humans. He wrote in the article, "This becomes especially obvious when taking into account that in the course of evolution such complex multi-gene determined anatomical and physiological structures were involved, such as the system of movement and orientation of the body in space, hand function, the structure of the skull, larynx and vocal cords and tongue." Bolstered in part by King and Wilson's discussion of the human and chimpanzee genome, Dmitri argued that destabilizing selection must have been at play, working through dramatic changes in gene expression. In his keynote address, he proclaimed that the huge number of changes in both body and behavior "involved not so much the structural as the regulatory elements of the genome." And those regulatory elements are mostly about gene expression patterns.

He thought the first major change was the transition by the *Australopithecines* to bipedalism—starting to stand upright. He argued that this involved not only the transformation of the entire motor system—both the nature of our bone structure and of our muscles—but also the emergence of important new brain abilities, particularly those involved in balancing upright. The mastery of this skill, he continued, then led to two new capabilities that were crucial in catalyzing further changes: the ability to see wider and farther and the freeing up of the forelimbs, which in time evolved into hands. Natural selection would clearly have strongly favored the emergence of these changes as they provided so many survival advantages. The acquisition of these new talents, he argued, then had a dramatic effect on the further development of the brain, pointing out that *Homo erectus*, which at the time was thought to have emerged about 1.3 million years ago,[7] had a brain a great deal larger than those of the *Australopithecines*.[8] So large had *Homo erectus*'s brain become that it was almost the size of our current *Homo sapiens* brain. This massive brain growth was accompanied by additional major changes in the body, such as the organs involved in the sensory functions and in speech—which included a substantial increase in the size of the larynx and a repositioning of the tongue—as well as the refinement of the motor skills of the forelimbs. This was vital, along with the emergence of better cognitive abilities, in allowing them to begin making tools. The interplay between the brain and body was at the core of his account. He wrote in the article, "We can say that if the body created the brain, and the individual mind generated by it, then the brain, in turn, is strongly influenced by the body functions." And that feedback loop had led to acceleration in the rate of change. He was keen to note that while the *Australopithecines* had evolved over the course of several million years, *Homo sapiens* had evolved into modern humans in less than two hundred thousand years.

Belyaev knew that many would think he was already stretching the limits of his destabilizing selection theory and what it might account for in our own evolutionary history, but Dmitri was not through yet. Never one to shirk what he saw as his duty as a scientist, he thought that this was an important enough issue to take some conceptual risks: time would tell whether he was on the mark. Dmitri next proposed that the combination of new talents he discussed had facilitated an intensely social way of life. These earliest humans organized themselves into larger social groups and developed many rituals, including religious practices, as well as making increasingly sophisticated artwork, such as the gorgeous cave paintings at Lascaux and Chauvet in France, making clothing, and developing more elaborate language. "The social environment," Dmitri said in his keynote speech, "created by man himself has become for him quite a new ecological milieu." In this speech he suggested, "Under these conditions, selection required from individuals some new properties: obedience to the requirements and traditions of the society, i.e., self-control in social behavior." These "new properties" destabilized the system, selecting for dramatic changes in behavior, which Belyaev thought likely occurred via changes in gene expression. This is where he makes a key connection to the process of domestication and self-domestication.

Humans who were better able to cope with the new stresses, to stay calm, cool, and collected rather than striking out in aggression, now had the selective advantage. "One can hardly doubt," Belyaev mused, "that the 'word' and its meaning has become for man an incomparably stronger stressful factor than a club blow for a Neanderthal man."[9] Calmer, cooler-headed members of communities were selected for, he proposed, with results similar to the effects of the artificial selection for tameness in the foxes. As with other domesticated species, this selection pressure led to lower levels of stress hormones, and it favored anything that prolonged our juvenile, more carefree, less aggressive stages of development. We also, like other domesticated species, can breed all year round. Essentially, we are domesticated, but in our case self-domesticated, primates. We sped up the process, Belyaev argued, domesticated ourselves even more quickly, because we, in turn, preferred tame partners as our mates.[10]

Primatologist Richard Wrangham has recently written about how just such a self-domestication process may be under way in another primate species, the bonobo (*Pan paniscus*), one of our closest evolutionary relatives. In 2012, he collaborated on a paper—"The Self-Domestication Hypothesis: Evolution of Bonobo Psychology Is Due to Selection against Aggression"—with a former PhD student of his, animal cognition specialist Brian Hare.[11]

Bonobos live a peaceful, one is tempted to say even enjoyable, life. They too have a fission-fusion society. Bonobo society is completely matriarchal, with females forming alliances amongst themselves. If a male has standing in a bonobo society, it is primarily because the females in the group permit it. Bonobos play all the time. They voluntarily share food, even with strangers. And sex is everywhere. But most sex isn't about copulation between a male and a fertile female. Homosexual sex among females is quite common, as is heterosexual sex between young and old, involving kissing, oral sex, and rubbing the genitals of a partner (either homosexual or heterosexual): primatologist Frans de Waal has quipped "bonobos behave as if they have read the Kama Sutra, performing every position and variation one can imagine."[12] Sex is the glue that bonds bonobo groups together. It is used as a greeting, it's a form of play, and it resolves conflicts that emerge. In this regard, bonobos are strikingly different from their close relatives, the chimpanzee.

Chimp society is patriarchal, males are violently dominant to females, they constantly fight one another to rise up the male hierarchy, and sex is about procreation. Males often form alliances, but unlike the female coalitions in bonobos, such alliances raid and viciously attack individuals in other groups. In bonobos, though intergroup interactions are sometimes nerve-wracking, in most instances they are peaceful gatherings, sometimes even involving copulations.

How could two such close genetic relatives have evolved along such different social trajectories? Wrangham and Hare became obsessed with finding an answer.

A molecular genetic comparison of the genome of chimpanzees and bonobos mapped onto an evolutionary tree suggests that they began to diverge from a common ancestor approximately two million years ago, at about the same time the Congo River was forming in Africa. The river split up the population of their common ancestors into two groupings, with those that evolved into bonobos living in a small area to the south of the Congo River, while those that evolved into today's chimpanzees lived north of the river and over a much larger area that stretched across west and central Africa.[13] Hare and Wrangham argue that by the luck of the draw, the bonobo lineage back then ended up with a huge advantage when it came to procuring food. Their territory boasted higher-quality plant foods. And what's more, they faced less competition for food. There were no gorillas where bonobos lived and so, unlike chimpanzees, they did not have to compete with their larger primate cousins for food.

In this world of relative plenty, with little competition over food, play, cooperation, and *tolerance of others* were advantageous. Bonobos who played during free time and cooperated with one another to obtain food, shelter, new friends, and sexual partners when play time was over, fared better than aggressive intolerant types. This selection for tameness led to changes in their bodies and behavior that are strikingly similar to the changes in the foxes.

Compared to chimpanzees, bonobos have more juvenile skeletal features, lower stress hormones levels, and altered brain chemistries. Like the tame foxes, bonobos also have a longer developmental period in which they rely on their mothers, show more variation in color (white color tufts and pink lips), and have smaller skulls, yet, remarkably, still have more gray matter in their brains devoted to areas linked with empathy than do chimps.[14] Hare and Wrangham go on to propose that over time, bonobo females may have selected the least aggressive, most friendly partners of the opposite sex as their mates. They may have self-domesticated themselves, in a process similar, but certainly different in the details, to that Belyaev outlined for human self-domestication.[15] Future work on whether bonobos have indeed undergone this process of self-domestication might, as Hare and Wrangham note, look at the role of gene expression and aggression, how neurobiological and hormonal differences affect aggression and tameness, and the intricate details of why behavior and morphology are so closely linked in both chimps and bonobos.

FOR YEARS DMITRI TOYED WITH HOW TO SET UP AN INDIRECT EXPERIMENT that would test his self-domestication hypothesis in humans. The idea would be to select a primate for tameness and see whether domestication followed. If it weren't so fraught with ethical issues, it would, given a great deal of time and research funding, he thought, be possible to perform the equivalent of the fox experiment with chimpanzees. Chimps and humans share a recent common ancestor, as foxes and dogs do. If an experiment, similar to what he and Lyudmila were doing with the foxes, selected the tamest chimps in every generation to mate with one another, how domesticated might they become? As a brilliant geneticist and evolutionary biologist he knew humans didn't evolve *from* chimps—we just share a common ancestor—and so he didn't think chimp domestication would replay human evolution per se, but instead might provide some hints on the role of domestication in our own evolutionary history.

Dmitri knew that performing such an experiment was out of

bounds, and he never took any concrete steps to explore the possibilities. But he did talk with friends and family about the idea. Pavel Borodin, of the rat domestication experiment, remembers a meeting where Belyaev brought up the chimpanzee idea. "We were rarely surprised by anything Dmitri said," Pavel says, "but this took our breath away." After discussing the idea for a bit, Pavel said, "Dmitri, do you understand what you're starting? Don't we have enough of our own problems . . . ? Do we really need to look at ourselves in the mirror?" Dmitri paused and told him, "You're right, absolutely right. *But* it is interesting, is it not?"[16]

Belyaev's son Nikolai recalls another occasion, when a colleague responded with shock to the very notion, saying, "It will take at least 200 years, so we will not know the results. Even if you are right, which is unlikely, what about the ethical issues?" Dmitri, who had little tolerance for shortsighted thinking, replied, "You don't see further than your nose. Certainly we wouldn't see the results, but other people would."[17] Then again, he hadn't expected to see results so quickly with the foxes, so who's to say how quickly the changes of domestication might also emerge in chimps? That is one question to which Dmitri would not be able to find the answer.

In the early winter of 1985 Belyaev was hospitalized with a severe case of pneumonia.[18] He was placed in the intensive care unit, and initially, he was so weak that physicians wouldn't even let his wife enter to see him. Only Dmitri's younger son Misha, by now a physician himself, was permitted to visit. But, very slowly, he began to recover. And as he did, he expressed one wish: to be well enough to be able to celebrate the fortieth anniversary of the defeat of the Germans in World War II, what Dmitri, like all in the Soviet Union, called the Great Patriotic War. He had never missed a Victory Day celebration before, and he had no intention of missing the one on May 9, 1985.

When Victory Day came, Dmitri gathered all the strength he had and walked up the steep flights of stairs to the hall where the celebration was taking place. When he entered the room, his friends and former brothers-in-arms—all of whom knew how sick he still was—gave him a standing ovation.[19] It was one of the last truly joyful moments he would experience.

With his condition persisting, he was advised to go to Moscow for specialist care, and there he was diagnosed with terminal lung cancer. The habitual smoking had finally caught up with him. His physicians wanted to get him back to Novosibirsk immediately, so he could spend as much time as possible with his loved ones. As an Academician—a full member of the Soviet Academy of Sciences—he was entitled to a flight by a special military aircraft, which they arranged. But when he learned that the flight would cost a small fortune, he had stopped the plan. No one, he believed, should have such a privilege above others. A regular flight back home would suffice.

For two months, he was still strong enough to communicate, but he was bedridden and frustrated that he could not continue with his work. "I need to work," he told one of his doctors, "but everyone is fussing around, restricting me, giving me handfuls of pills."[20] He was allowed to stay in his home, with oxygen tanks brought in to support his failing lungs, and the close-knit community of the Institute of Cytology and Genetics gathered around him.

As the end neared, Dmitri arranged to give one last interview to the press. He used the opportunity to share his vision of the future. "Within a couple of decades," he told the reporter, "humans will be able to fully study our planet to the very core . . . exploit the near-earth space . . . work for a long period of time in zero gravity, and create around the earth, at its orbit, enclosed ecosystems. . . . All aspects of human activity will be successfully improved through . . . automation. We will see fifth and maybe sixth generation computers. These will be talking, thinking, and self-innovating machines. Personal computers, robots, and communication systems will be widely employed." That much he felt certain about. "But what will become of humans," he added, "I do not know."

When the reporter followed up by asking him what he *wished* for mankind in the twenty-first century, Belyaev replied, "Be kind and socially responsible, strive for mutual agreement with all people, live in peace, carry a full and sincere responsibility for our 'younger brothers'—all living creatures on earth. We should never forget that we are just part of nature, and we should live in harmony with nature when we study its laws and use this knowledge to our service."[21] Just as he had done.

On November 14, 1985, Dmitri Belyaev died, surrounded by loving friends and family, and secure in the knowledge that his life's work would continue. He had trained the vice director of the Institute, Vladimir Shumny, well, and he was confident Shumny would take charge smoothly. Of course he knew that Lyudmila and the fox team would keep the domestication experiment going, and he was sure they would make many wonderful new discoveries.

He did have one regret. "He wanted to write a book," Lyudmila says. "His greatest desire was to write a book on domestication . . . and the book was supposed to be popular . . . He wanted to tell stories, to tell the layman, anyone . . . what processes had underlain domestication," Lyudmila continues, "why we have these animals living around us, why they are the way they are." Belyaev had talked to Lyudmila and others about his dream of writing this book many times, but he had been most adamant about it when he learned of one special story about Pushinka. Many years earlier, Lyudmila had acted out, in vivid terms for Dmitri, how Pushinka, immediately after giving birth to her pups, had brought them to her and placed them at Lyudmila's feet. "When I narrated that nice story to Belyaev," she says, "he was so surprised, he was so perplexed, he was so intrigued, that he said that we should write a popular book to tell people . . . to make people understand domesticated animals . . . why [and how] they behave differently from their wild [ancestors]." He even had a name for the book: *Man Is Making a New Friend*.

The day of Belyaev's funeral saw sleet, snow, and rain. Looking back on the ceremony, Dmitri's family, friends, and colleagues have mixed feelings. All agree that the funeral, and the other ceremonies that were linked to it, garnered the attention that a man of Belyaev's stature merited. The crowd was huge: a mixture of fellow scientists, staff from the Institute of Cytology and Genetics and many other institutes at Akademgorodok, family, friends, and former comrades from the Great Patriotic War. And then there were the dignitaries, both political and scientific, who came from as far away as Moscow. Many of them had never met Dmitri Belyaev, but they dominated the podium, giving the sort of laudatory eulogies that VIPs specialize in.

Respectful as all the speeches were, the staged, bureaucratic nature of the funeral left no time for friends and family to share their thoughts: they were not given any time to stand up and give *their personal* eulogies. That stung, and the anger and disappointment has stayed with them to this day. "I wanted to speak up," Lyudmila says, but protocol just didn't allow it. She and others could stand and watch. But after all was said and done, something happened that refreshed their spirits. A woman approached Lyudmila and those around her. The woman was weeping as she said, "You don't know to whom you are saying good-bye for good today." Lyudmila and the others were taken aback. "What do you mean we don't know him?" Lyudmila said. "We have known him for more than 20 years!" To which the woman replied, "Perhaps you have known him for 20 years, but, you don't know what sort of man that man was." And then she told a story that no one there ever forgot.

She had been a bank teller. Years earlier she had suffered from severe pain in her legs. One day, Belyaev was in the bank when he overheard a conversation between this woman and a colleague of hers. The teller was describing the pain in her legs, and how she wasn't sure how much longer she could even keep her job in the face of this daily pain. What would happen to her and her family then? Her colleague told her that she needed to visit a doctor immediately. "I have been to all the doctors," the teller replied, "but they are not helpful. I want to be put in hospital, but they say there are not enough beds. I don't know what to do: no one does." Belyaev listened, finished the work that brought him to the bank, and left. Two days later the woman received a call while at work. The voice on the other end told her that there was a room available for her in the hospital and that she should proceed there as soon as possible.

Shocked, the teller said, "That is impossible, I have been told many times there are no beds for me." That might be, said the caller, but we have been contacted by Academician Belyaev and he has asked that we remedy this situation. The woman went to the hospital, underwent a successful series of surgeries, and soon returned to her teller job pain free. Belyaev, as was his nature, never mentioned a word of this to anyone.

Fig. 1. A domesticated fox enjoying the summer shade at the fox farm outside of Novosibirsk. Credit: Irena Muchamedshina

Fig. 2. An inquisitive domesticated fox peering out from behind vegetation. Credit: Irena Muchamedshina

Fig. 3. A domesticated fox relaxing. While Siberia is brutally cold during the winter, it gets quite hot during the summer. Credit: Irena Pivovarova

Fig. 4. A domesticated fox pup playing. Credit: Anastasia Kharlamova

Fig. 5. A domesticated fox rests its head on the shoulder of one of the members of the fox team. The bond between tame foxes and humans emerged early in the fox farm experiment. Credit: Irena Pivovarova

Fig. 6. Left to Right, Lyudmila Trut, Aubrey Manning, Dmitri Belyaev, Galina Kiseleva, all sitting on a bench in front of one of the tame foxes. Years earlier, Lyudmila was seated on this bench when Pushinka barked at an intruder. Credit: Aubrey Manning

Fig. 7. Two domesticated foxes, one with a balloon toy in its mouth. These foxes play with almost any object they can get into their mouths. Credit: Anna Kukekova

Fig. 8. Two domesticated foxes playing in the winter snow. Credit: Aaron Dugatkin

Fig. 9. Two workers carrying domesticated foxes at the fox farm. Winter days at the fox farm can be short and frigid. Credit: Aaron Dugatkin

Fig. 10. Two domesticated fox pups being taken for a walk by Irena Muchamedshina. On occasion, the tame foxes are walked around on a leash and act in an astonishingly dog-like manner. Credit: Anastasia Kharlamova

Fig. 11. A gorgeous domesticated fox. Credit: Institute of Cytology and Genetics

Fig. 12. Lyudmila Trut with one of her beloved domesticated foxes. Credit: Vasily Kovaly

Fig. 13. A domesticated fox pup out for a leisurely stroll. Credit: Irena Muchamedshina

Fig. 14. Tatjana Semenova hugs two domesticated fox pups. Credit: Vladimir Novikov

Fig. 15. Domesticated fox pups with Irena Muchamedshina. Credit: Irena Muchamedshina

Fig. 16. Three adorable domesticated fox pups sitting together in the field.
Credit: Irena Pivovarova

8

An SOS

The year of Dmitri Belyaev's death, 1985, ushered in a period of great tumult in the Soviet Union. The top-down communist system was entering its death throes. When Mikhail Gorbachev became secretary general in March, he implemented the policies known as glasnost (openness) and perestroika (restructuring), intended to make the Soviet government more transparent and the economy more efficient. Instead, they sent the system into shock. The economic restructuring Gorbachev instituted led to massive shortages of goods from oil to bread and butter, and strict rationing was instituted. The Soviet people found themselves waiting in long lines for even the most basic essentials.

For a time, the scientific work of the Institute of Cytology and Genetics was protected from the economic upheaval, and Lyudmila was able to continue operations as normal at the fox farm. The new director, Vladimir Shumny, appreciated the importance of the fox experiment and made sure it was as well funded as possible. Lyudmila took over all responsibility for the running of the experiment. She missed Dmitri terribly, and thought about him every day as she arrived at her office to pore through some new data about the foxes or as she checked in on the new generation of pups, whom he would have loved visiting. Working hard to keep his spirit of scientific exploration alive with her team at the farm, she launched several important new studies.

The number of pups born each year who displayed most or all of the elite traits increased at a quickening pace in the 1980s, so that by mid-decade, out of about 700 foxes on the farm, between 70% to 80% were in the elite category. Additional changes in both their looks and their behavior also emerged. In addition to more of the foxes having curly tails, their tails were also becoming bushier. Many of the foxes also began to vocalize in an odd new way, making a high-pitched "haaaaaw, haaaaaw, haw, haw, haw" sound when people approached them. Lyudmila thought it sounded like they were laughing and called it the "ha ha" vocalization. She was also now sure that the foxes' anatomy was changing. There was no longer any question that the snouts of many of the foxes born in these generations were slightly shorter and more rounded, and their heads seemed to be somewhat smaller also. These changes in anatomy were significant enough now that Lyudmila decided the team should make measurements, to compare the snouts and heads of the elites to those of the control foxes.

Reading about the latest techniques in anatomy research, Lyudmila learned that ideally, they should take X-rays of the foxes' heads and then make their measurements from those. But she didn't have access to an X-ray machine, and though so far her budget for running the experiment hadn't been slashed, she couldn't allocate the resources for such an expensive purchase. So she and her team would have to do the work the old fashioned way, making direct measurements of the foxes. This was difficult and time-consuming work, which required the workers to help out again, holding the foxes still while Lyudmila and her research team measured the height and width of their skulls and the width and the shape of their snouts. Their arduous work paid off. They found that the skulls of tame foxes were significantly smaller than those of the control foxes, and the differences in the snouts were somewhat more pronounced, with those of the tame foxes being in fact considerably rounder and shorter than those of the control foxes. These same changes were involved in the evolution of the dog from the wolf: the skulls of adult dogs are smaller than those in adult wolves and their muzzles are wider and rounder.[1] These changes in anatomy were another way in which dogs, and now the tame foxes, retained more juvenile features as they matured. When Lyudmila had compiled all of the data and saw the stark differences, she thought to herself, *Dmitri would have been pleased*. These changes added to the domestication package; the tamest foxes were now displaying so many of the types of transformation seen in domesticated species.

Another study Lyudmila launched looked deeper into the changes in the level of stress hormones in the tame foxes. Instead of just measuring hormone levels in the foxes as they had done before, this time she and her colleagues Irena Plyusnina and Irena Oskina would experimentally manipulate the levels in order to see whether behavioral changes might result. They already knew that, compared to control foxes, the tame foxes had a significantly lower level of stress hormones after the point, about 45 days into development, when their production spikes up in wild foxes. They had subsequently found that in the aggressive foxes, the spike up in stress hormone levels was significantly higher than in the control foxes. Now, to produce definitive evidence that the behavioral differences between the two lines were due primarily to these different levels of stress hormones, Lyudmila decided to conduct a study to discover whether, if the level of stress hormones in the aggressive foxes was lowered, they would act tamer. It was now possible to experimentally block the spike up in production of the hormones in the aggressive foxes by feeding them a capsule filled with chloditane, a chemical that stops the production of some stress hormones.[2] Lyudmila selected a number of pups whose mothers and fathers were aggressive, and Irena fed them the capsules starting just a little before the 45 day mark. Another group of pups from aggressive moms and dads served as a control group, and they were fed a capsule full of oil. The results were striking; the pups given the chloditane, who did not produce a surge of stress hormones, acted more like tame pups, while those fed the oil developed into normal aggressive adults.[3]

Lyudmila then decided to perform a similar experiment with serotonin levels, which she had found were so much higher in the tame foxes.[4] In this case, starting when they were 45 days old, she would *increase* the amount of serotonin in the systems of one group of pups born to aggressive parents, while pups of aggressive parents in one control group would be given no injections and those in another control group would be injected with a solution of water and salt. Again, the results were crystal clear. The pups in both control groups developed into aggressive adults, while the pups given the extra serotonin did not: they acted more like tame foxes.[5]

From that day in May 1967 when Belyaev had called Lyudmila into his office to share his tour-de-force new idea, changes in hormone levels had been at the core of his theory of destabilizing selection. The results from this new manipulative work on stress hormones and serotonin fit in beautifully.

BY THE LATE 1980S, THE FOX domestication experiment was approaching age thirty, making it one of the longer ongoing experiments in animal behavior ever conducted. Then suddenly, it seemed that it would come to an abrupt and tragic end. The upheaval in the Soviet economy had intensified as the decade proceeded, and the Union had begun to crumble. So dire did the prospects of the fox farm become that Lyudmila and her team found themselves fighting desperately to keep the foxes alive.

In 1987, protests against Soviet control broke out in the Baltic republics of Latvia and Estonia and spread throughout the Union. In 1989, the pro-democracy Solidarity movement in Poland forced the Soviet government to allow free elections, and on November 9 of that same year, with massive crowds of pro-democracy protestors marching in East Berlin, the guards at the Berlin Wall stood down and crowds of revelers climbed atop the wall cheering. On October 3, 1990, East and West Germany were official reunited. In early December 1991, the Supreme Soviet renounced the treaty that had formally established the Union, and by December 21, fourteen of the fifteen Soviet republics had formerly withdrawn from the Union, and eleven had joined together creating the Commonwealth of Independent States. On December 25, Mikhail Gorbachev stepped down from the presidency, and the Soviet flag flew over the Kremlin for the last time.

The top-down command and control system that had overseen every aspect of life in the Union fell into chaos, and funding to agencies and institutes of all kinds either ceased or was drastically cut. The budget of every institute in Akademgorodok was slashed. Most labs still had some stores of equipment and materials with which to continue at least some research, but the fox farm faced an immediate crisis. Lyudmila was left with virtually no funds to pay the workers and little to buy food for the foxes. At this time, the population of foxes still sat at around 700 animals, and their food alone was a considerable expense.

She had to announce to the staff, who cared so much for the foxes and had been so dedicated to helping with the research, that she could no longer afford to pay them. Some stayed on anyway. They couldn't bear to leave Lyudmila or their friends the foxes. Lyudmila asked those who felt they had no choice but to look for other work to please come back after she had somehow found funding again. "We told them," she recalls, "when we are more or less okay again, please come back, *we need you*." In the meantime, caring for the foxes and fighting to keep them alive became her all-consuming passion.

The director of the Institute of Cytology and Genetics squeezed all the money he could out of his budget to send Lyudmila. The fox experiment was the Institute's greatest achievement. As Lyudmila says, it "became the 'business card' of the Institute," spreading the word of the excellence of work at the Institute to the world genetics community. Lyudmila sent a grant for funding to the Siberian Academy of Sciences, and recognizing the importance of the experiment, the Academy provided some money. The new funding allowed Lyudmila to feed the foxes, but research work had to be put on hold. Then, in 1998, the bottom dropped out of the Russian economy. A severe economic crisis led to the devaluation of the ruble on the world market and then, in August, Russia defaulted on its public debt,[6] which caused a severe currency shortage. Funding for government-run enterprises of all kinds dried up completely, and Lyudmila was left with virtually no money at all coming in for the fox farm. She and all those at the farm who loved the foxes so much now faced the horrifying prospect that they might not be able to keep the foxes alive.

The farm had some store of food, and Lyudmila had squirreled away a little money over the years from grants that allowed her to continue to buy some food and medicines that were critical to stopping the spread of disease, such as the fox version of hepatitis and any number of gut parasites that infect them. When that dried up she and a few colleagues at the Institute did everything they could think of to raise funds to buy as much food as they could afford. But it wasn't nearly enough to keep the foxes fully fed and they began to lose weight. Lyudmila was so desperate to somehow stop the foxes—her foxes—from starving that she went out to the roads around the farm and the Institute and stopped cars, asking people for money or any kind of food at all they could give.

Lyudmila decided she had to make a plea about the foxes and the dire straits they were in. She sat down to write an article all about the experiment, and to send an SOS to both the scientific community and the broad public for support. Perhaps they would send help. "Forty years into our unique lifelong experiment," Lyudmila wrote, "we believe that Dmitry Belyaev would be pleased with its progress. . . . Before our eyes, 'the Beast' has turned into 'Beauty.'"[7] She described the full cascade of changes they had seen in the foxes, explaining what loveable and loyal animals they had become. "I have raised several fox pups in domestic conditions," Lyudmila wrote. "They have shown themselves to be good-tempered creatures . . . as devoted as dogs but as independent as cats, capable of forming deep-rooted pair bonds with human beings—mutual bonds." You know these animals, Lyudmila was telling her audience, they're just like the pets you have in your homes, the pets that you love and that your kids love. She also called out the many avenues of continuing investigation. Analysis of the foxes' genomes had yet to be done, they still needed a much deeper understanding of how some of the foxes had reproduced more than once a year, they had started to hear new vocalizations in the tame foxes and wanted to know why, and they had only just started working on cognition in these special animals. And at the broadest level, though they had been at this now for forty years, that was just the blink of an eye in evolution—how far could they take domestication of the tamed fox, given more time?

She closed with a forthright assessment of how stark the situation had become, but she didn't actually ask for support. "For the first time in 40 years, the future of our domestication experiment is in doubt," she wrote. After describing the dire plight, she ended by announcing that she hoped to someday make elite pups available for people to adopt as house pets.

She sent the article off to one of the leading popular science magazines in the US, *American Scientist,* along with a number of photographs of the foxes showing how dog-like and affectionate they were, including one of Dmitri sitting with a group of pups playing around his feet and jumping up to lick his hands. She hoped the editors would understand the value of keeping the foxes alive and print it soon.

Despite all her efforts, as winter approached, foxes began to die. Some were felled by disease, but most died of starvation. She and her research team and the workers who had stayed on to keep the cages clean and offer any care at all they could, agonized as the population dwindled. To her horror, Lyudmila was faced with the excruciating proposition that the only way she could find funds to prevent the wholesale death of the foxes was to begin sacrificing some of them, in order to sell their pelts. She instructed that they be euthanized at the farm, to die in a peaceful manner, rather than be taken away. She chose mostly foxes from the aggressive and control populations for this, selecting those who were in the worst health and closest to death, shielding the tame animals from this awful fate to the extent possible. Making these selections was the hardest thing Lyudmila had ever had to do, and she finds it extremely difficult to discuss this horrible time still today. Some of the caretakers and researchers were so deeply traumatized by this turn of events that they needed counseling, and one worker completely broke down and had to be treated in a psychiatric ward.

By early 1999, only 100 tame females and thirty tame males were still alive, and even fewer aggressive and control foxes. Lyudmila now felt the only hope was that her article would appear in *American Scientist* and people would be moved to help. Day after agonizing day passed with no word, until one day she was elated to discover that she had received word from the magazine's editor. With great trepidation, she read it, and the news was good—the article had been accepted.

Titled *Early Canid Domestication: The Farm-Fox Experiment*, the article appeared in the March/April issue in 1999, displaying several of the photos Lyudmila had sent, including one of Dmitri with the pups, and one of a researcher holding a fox that was licking her face. When later she got the word that longtime *New York Times* science writer Malcom Browne had written an article in the *Times* telling the story of the foxes and referring to her plea, she felt a rush of hope. But Lyudmila worried that perhaps she was simply dreaming, grasping at straws. Would people respond? Would anyone actually send support? She worried, she recalls, "Maybe I was wrong about how others would feel."

She wasn't wrong. The response was heartwarming. Animal lovers all around the world heard her call, and letters immediately flooded in. "I was alarmed by your final paragraph," one man wrote. "Is it possible for a private American citizen to make a direct contribution to your center? I cannot afford very much, but I would be willing to invest a small amount to make a statement of my support."[8] Another man, an offshore oil driller, wrote, "I cannot afford a lot but I can help . . . Please send me a way to donate."[9] Some people sent a few dollars, a few sent $10,000 or $20,000. Lyudmila was able to buy the foxes all the food and medicine they needed again, and to bring back some of the caretakers. The foxes, and the experiment, were saved.

The scientific community also rallied. The story of the foxes was the buzz at scientific conferences all over the world, the hot topic of discussion during coffee breaks between paper sessions. Geneticists and animal behaviorists realized that this extraordinary line of domesticated foxes could provide important clues not only about the genetics of domestication but also about the link between genes and behavior. There were so many potential avenues for investigation. The genome of the foxes could be sequenced, which the Institute of Cytology and Genetics did not yet have the technology or funding to do. More studies could be done to analyze the changes going on in their hormone production and the genetic causes of those changes. A new boom in the study of animal cognition and the nature of the animal mind was underway, and the cognitive abilities of the foxes would be a great topic for examination. Lyudmila began to receive inquiries from scientists abroad, and she opened the arms of the fox farm to them.

One of the first of many scientists to contact Lyudmila to explore doing research with her on the foxes was Russian-born geneticist Anna Kukekova, who had received her PhD from the University of St. Petersburg and then taken a position at Cornell University, studying the molecular genetics of dogs. Anna had first contacted Lyudmila in the early 1990s, as an undergraduate, hoping to do some work with the fox team, but the Institute was in the midst of the first economic pains then and couldn't bring her in.

Anna's strongest interest had always been in dogs and their relatives. When she was twelve, she had joined the Club of Young Zoologists at the Leningrad Zoo and when asked to choose a favorite animal to learn about, she selected the Australian dingo, because she was curious to understand what made these wild dogs behave differently from other dogs. Her passion had carried through to her graduate days, when, even as swamped as she was with her research on bacteria and viruses, she managed to find time to work as a dog trainer a few days per week.

After she got her degree, she searched for jobs in the emerging field of dog genetics. At the time, only a handful of labs were working on the dog genome, and Anna wrote to many of them. Greg Acland's lab at Cornell had recently received a hefty grant, and he made her an offer. She left Russia for the pastoral hills of Ithaca, New York, in 1999.

It was an auspicious time to be diving into molecular genetics. The prior decade had been a watershed period for discovery in genetics, with powerful new tools for analysis of genes introduced and a rush of important findings. In 1983, scientists had mapped out the location of the first disease-causing genes in humans, the genes linked to Huntington's disease, which were located on chromosome 4 in humans. That same year chemist Kary Mullis invented the technology for rapid replication of fragments of DNA known as the Polymerase Chain Reaction (PCR), which would win him a Nobel Prize ten years later, for revolutionizing how quickly and how accurately genes can be mapped. By 1990, a key mutation in the gene associated with cystic fibrosis had been identified, and molecular genetics was hot on the trail of understanding how tumor suppressor genes went awry and led to breast cancer sites. The year 1990 also saw the start of the Human Genome Project, a monumental worldwide collaboration.

The first complete genome of any free-living species mapped was that of the bacteria *Haemophilus influenzae*, which despite its name does not cause flu, but does cause severe cold symptoms, especially in young children. Researchers found that its genetic code comprised 1.8 million letters, which suggested that the genetic codes of more complex species could be staggeringly long. The next year, the genome of the first fungus, popularly known as baker's yeast because it is used to make bread dough rise, was mapped. Then, in 1996, to the delight of many, and the sheer horror of others, who thought science was now venturing into areas it had no business in, developmental biologist Ian Wilmut and his team at the Roslin Institute in Scotland took a mammary cell from one sheep, implanted it into the emptied-out egg of another sheep, and implanted the egg into yet a third sheep. On July 5, 1996, that sheep gave birth to the first sheep's clone—6LL3, soon named Dolly at the suggestion of a Dolly Parton fan who helped deliver her. Princeton biologist Lee Silver summed up both the delight and the fear: "It's unbelievable. It basically means that there are no limits. It means all of science fiction is true. They said it could never be done and now here it is, done before the year 2000."[10]

The first complete sequence of an animal—the nematode *C. elegans*, a workhorse of medical genetics–was released in 1998, comprising 100 million letters of code. Then in 1999, less than five decades after Watson, Crick, and Rosalind Franklin had solved the puzzle of the structure of DNA, and nine years after the Human Genome Project was initiated, scientists in England, the United States, Japan, Germany, France, and China published the map of the first of our own 23 chromosomes. Human chromosome 22 was mapped out first because it was relatively small, and it had been linked to numerous diseases. Just two years later, a first draft of the human genome was published in rival papers in the world's two leading journals, *Science* and *Nature*: one paper came from the Human Genome Project team and the other from Craig Venter's team Celera Genomics. Francis Collins at the National Institutes of Health predicted this would eventually lead to "individual preventive medicine." Two years later, the project was declared virtually complete, mapping out, letter by letter—about 3.2 billion of them—99% of our genes. Many compared it to the moon landing in terms of a triumph of human endeavor.[11]

In late fall of 2001, just as the first draft of the human genome was being released, Anna learned about Lyudmila's *American Scientist* article and the foxes' dire straits. She looked up all the articles that had come out over time about the experiment to learn more about the work that had been done with them since she had last heard about the experiment. Discovering that no gene sequencing work had yet been done with foxes, she wondered whether the tools she was using in mapping the dog genome could be modified to map the fox genome. Perhaps if she began to map the genome of the elite foxes, then one day—maybe even in just a few years—comparing it to the dog genome would produce important information. And the questions she could address were endless, given how little was known about the tame foxes' genome.

Sequencing an individual gene, let alone being able to sequence a significant chunk of the entire genome of the elite foxes, was work that Lyudmila had thought would simply not be possible for her, or anyone, to perform for quite some time. To be able to then also compare the genome to that of the dog was a dream. Dog genomics was a new area of study, and very few researchers were trained in it at the time. But luckily Anna was one of them and she wanted to help usher Lyudmila and the foxes into this brave new world of discovery.

Anna proposed to her Cornell postdoctoral mentor Greg Acland that she call Lyudmila when she was over in Russia to spend the 2002 New Year's holiday with her mother and grandmother, and see whether Lyudmila would agree to work with her on the project. Greg thought it was a great idea. So, shortly after Anna arrived back home in Moscow, she called Lyudmila, who was thrilled with the idea. Anna had assumed that if Lyudmila was amenable, then once she returned to Cornell, she and Greg would follow up with Lyudmila to work out the details. But when Lyudmila asked her what the first step should be, and Anna told her that it would be to get blood samples from the foxes, Lyudmila suggested that she fly to Novosibirsk—now. Lyudmila had not led the fox experiment so successfully for forty-plus years by letting opportunities pass her by.

Anna was stunned. Get started? Now? She had expected back-and-forth discussions for months. But Anna also knew how to capitalize on an opportunity. There was one catch, though. Anna would need at least 300 vials for the blood, but because such equipment was rare and expensive in Russia at the time, the Institute of Cytology and Genetics didn't have them. She told Lyudmila she'd somehow get ahold of them. Calling old colleagues at the laboratory where she had done work at the University of St. Petersburg, Anna had the vials in hand in a couple of days. She flew to Novosibirsk on January 4.

Things continued to move at lightning speed. Within moments of Anna arriving in Lyudmila's office at the Institute, Lyudmila said to her, "We do not have much time, let's go to the farm." Anna vividly remembers how astonished she was when she met the elite foxes. "To say that I was amazed interacting with the tame foxes is to say nothing," she recalls. "I was amazed by the strength of the desire of these foxes to interact with humans." She put her emotions on hold, though. She had to get down to business right away, preparing for the sampling procedure. Ideally, she should get blood samples for the molecular genetic analyses from three generations of foxes, and Lyudmila immediately assigned two people on the fox team the task of going through their gigantic genealogical database to identify which foxes to draw blood from. So efficient was the fox farm team that by the next morning, when Anna arrived at the Institute at 9:00 a.m., the list of foxes was all ready for her.

Lyudmila had also made arrangements to do the sampling at optimal speed. They had only a couple of days to do all of the work, and in the bitter winter cold, it couldn't be done in the unheated sheds. The foxes would have to be brought inside. So Lyudmila organized an assembly line of ten of the caretakers, mostly women, to help take the foxes from their pens and bring them into one of the houses at the farm for drawing the blood. The pace of activity was intense, and when one worker slipped and broke his arm, he told the others to keep right on going and not concern themselves about him. It was a team effort par excellence. Anna was moved by the commitment of the workers. "It was an experience to meet these ladies," she says, to see their "deep passion for the animals. They reminded me of some old zookeepers in Leningrad's zoo from my childhood."

By sunset they had collected samples from about 100 foxes. The next day, they did the same again. "Lyudmila brought a cake to the farm for the animal keepers," Anna says, "to thank them at least a little bit for all the extra work that they did for us."

Anna didn't have the necessary permits for bringing blood samples into the US from overseas because she had not expected to get them on this trip. Fortunately, she didn't actually need the blood per se, just the genetic material in it. So when she stopped in St. Petersburg on the trip back to Cornell, she turned again to her friends at the university and they agreed to extract the DNA from the samples even though they had only five days before she flew home. Her friends rallied and they finished the job in just three. They understood how important the analysis of the DNA might be.

The drive to isolate genes linked to fox domestication was on.

9

Clever as a Fox

The opportunity to collaborate on studies with the domesticated foxes was of great interest not only to geneticists like Anna Kukekova, but to specialists in animal behavior as well. Just as Aubrey Manning was captivated by the early reports in Western scientific journals about the experiment when organizing his 1971 Ethology meeting in Edinburgh, in the 1990s a new generation of animal behavior researchers recognized how valuable the fox findings were to their work, and how important new studies conducted with the foxes could be. An explosion of new work was focusing on animals' cognitive abilities and the kinds of learning they are capable of. The foxes presented a golden opportunity for exploring differences in cognition between domesticated animals of a species and their wild relatives.

Lyudmila and Dmitri had agreed that the genetic changes leading to the foxes' domestication must have primed their brains for learning how to be more sociable with people. Pushinka had learned to show Lyudmila special loyalty, and she also believed Pushinka had likely demonstrated rudimentary reasoning ability. And Pushinka's clever deception in playing dead in order to catch the crow seemed to show strategic forethought. But Lyudmila did not have any expertise in studying animal cognition, and had not launched any studies to test the foxes' thinking skills.

Getting inside of an animal's mind is tricky. Any dog owner who has watched a dog gingerly grasp a rawhide bone in its mouth and proceed to a corner of the room, or behind a chair, pawing at the floor and then gently placing the bone down as if burying it has wondered what exactly is going on in his pet's brain. Is his little terrier or beagle playacting? Is he or she engaging in the equivalent of a child's imagined tea party or fire truck rescue? Or has the dog perhaps learned, smartly, that stowing the bone away for lean times ahead might be a good idea? When cats pounce at one another from behind a door are they creating scenes in their minds of masterful hunting feats, and are they envisioning themselves eluding a fearsome predator as they race through a room? Maybe instead, our pets are simply acting on instinct, as Charles Darwin surmised about the dog he observed turn round and round on a carpet thirteen times before settling down for sleep.

What exactly is the nature of animals' mental lives? We don't really know. The most difficult questions to answer about animal behavior have been those about the nature of animal minds and emotion. Darwin had conjectured that animal cognition and emotion are on a continuum with that of humans. But as researchers made more discoveries during the twentieth century about how genetically programmed animal behavior can be—such as Konrad Lorenz's demonstration that Greylag goslings would mistake a rubber ball for their mother during the imprinting period—they were extremely careful not to anthropomorphize and project human-like thoughts onto animals. Jane Goodall's assertions about chimpanzees had kicked up such a storm over inferences about animals' inner lives that the bar of proof was now set very high. But Goodall's observations, along with the observations of other animal behaviorists, had also stirred up interest in finding new ways to probe into the nature of the animal mind.

Many ethologists who undertook this work, such as Bernd Heinrich and Gavin Hunt, followed in the tradition of Lyudmila's mentor Leonid Krushinksy, as well as Nobel Prize winner Nikolaas Tinbergen, studying animals in the wild. A number of fascinating studies revealed that animals besides primates make use of tools. The crows of New Caledonia (*Corvus moneduloides*) are master toolmakers in the avian world.[1] These birds extract insects from under tree bark using tools they construct from twigs and leaves. They thrust the tools into cracks in the bark, and when the target insects react defensively and grab the tool, the birds pull the tools back and either gobble up the bugs or feed them to hungry offspring. They learn to make these tools during the first year or two of life, as they progress from apprentices, watching skilled toolmakers tinkering with their creations, to master toolmakers themselves. They begin by making the simplest tools, which are twigs that they strip leaves and side bits from, so they're nice and smooth for probing. Eventually they figure out how to fashion more complex tools, such as twigs with hooks at the end. To do this, they select a small branch that is forked into two slender branches, and they bite one of the branches off right above the base of the fork, so that the remaining branch now has a small "v" at the end, as when the wishbone of a chicken is snapped and one side is much shorter than the other. They then do a kind of whittling of the v with their bills to sharpen it.

These crows make tools from the barbed edges of leaves of the screw pine, tapering the leaves toward the end, akin to the tip of a spear, and then using these as probes to poke about for food. Researchers who have studied these birds in the lab discovered that they would also use novel objects like cardboard and aluminum to fashion tools, which led researchers to install a series of "crow cams" in their natural habitat in New Caledonia to see if they display such ingenuity in the wild. This produced footage of the birds sculpting molted feathers and dry grass into tools as well. The footage also showed that, on occasion, the birds even use their tools to get at lizards, an especially juicy and protein-rich treat. Remarkably, they also safeguard their best and favorite tools for reuse.[2]

Why New Caledonian crows show this dramatic tool-making expertise and other species do not is a subject of much discussion. Researchers have searched for what factors are operating in crows, but absent in other bird species that don't make tools, to find an answer. The working hypothesis is that a combination of conditions facilitated their development of this ability. Low competition from other birds for food and low predation rates are thought to have afforded them more time to experiment with tools, and the relatively long developmental period of these crows offers youngsters ample opportunity to learn the skills from their parents and other adults.

In addition to the studies on animal learning per se, a good deal of work has focused on their memory abilities, and some astonishing findings have been made. In the animal world, few can match jays, a member of the corvid family that includes ravens and crows, for feats of memory. While some jay species don't store much food for times of trouble, other species can remember the location of 6,000 to 11,000 seeds that they have stored over the course of nine months. This ability is linked to the very large size of the hippocampal area of their brains.[3] Western scrub jays take bird smarts up a notch. They not only recall where they have stored vast numbers of food items, but they remember who was watching them when they cached their food, and when they are watched will later dig up and hide the food elsewhere, presumably to protect their stored food from being stolen.[4]

Some species may have a rudimentary ability to understand numerosity. Chimps can recognize that there are more yummy pieces of sweet banana on one plate than another. Dogs accustomed to receiving a certain number of treats will make clear they expect the usual number if shorted. They will also be noticeably disturbed if treats are distributed inequitably to them and another dog gets more. Desert ants, who have few cues from their barren environment to help them find their way back home, can gauge how many steps they have taken from their nests on foraging journeys. Collecting a sample of ants that had gone out on a foraging party, animal behaviorists worked out a way to attach tiny stilts to the ants' legs, which made their legs 50% longer. Returning them to the foraging grounds and observing them make their way back to their nest, the researchers found that the ants walked 50% farther than they should have, at which point they stopped and began searching for their nest. They had overshot it by just the distance that the extra length of their strides accounted for, which can most reasonably be explained by them having kept track of how many steps they had taken.[5]

During this boom of research, the study of reasoning ability also advanced considerably. A number of researchers began asserting anew that some nonhuman animals demonstrate reasoning skills. The strongest claims were made, of course, for primates. The notion that primates have human-like reasoning ability actually went back to the early twentieth century. German researcher Wolfgang Kohler observed apes while he was head of the Prussian Academy of Sciences primate research station in the Canary Islands in the 1910s, and he wrote about how creative they were in solving many problems. He had watched them put wooden crates one on top of the other in order to climb up on them to reach bunches of bananas, and also use long sticks to whack them down from trees. He described these feats in an influential book, *The Mentality of Apes*, first published in Germany in 1917, in which he argued that the apes had clearly applied reasoning skills to accomplish them. His work had fallen out of favor in subsequent decades as so much work focused on conditioning and instinct alone in explaining animal behavior. But the observations of Jane Goodall, Diane Fossey, and others of chimps and gorillas, followed up by a new generation of primatologists, including Frans de Waal, Dorothy Cheney, Robert Seyfarth, and Barbara Smuts, who made observations of the complex social lives of bonobos and other primates in the wild and in the laboratory, brought it back into vogue.

One particularly fruitful vein of research in this area has been the study of animals' social cognition, meaning their ability to evaluate the social situation they are in, such as with chimpanzees foraging in a group in the forest, or a group of dogs let out to play in a dog run. Researchers study how animals respond to cues from one another, or respond to cues from other animals, such as the way dogs are so adept at reading their owner's moods. This was the area of work to which the tame foxes were to make a major contribution.

One of the leading contributors to this research on animal social cognition traveled to Akademgorodok to conduct an intriguing study with the foxes. Brian Hare was still working on getting his PhD at the time, under the supervision of Richard Wrangham, with whom he went on to write the papers about self-domestication in bonobos. Brian's specialty was in comparing social cognition abilities across animal species, and he was focusing on studying dogs and primates. He was particularly interested in understanding how their social skills had evolved.[6]

There was no question from Brian's own work, and the work of others, that non-human primates, like chimpanzees and baboons, showed sophisticated types of social cognition. This was seen, for example, in the way primates groom one another.[7] Researchers sweating in the sweltering heat of Africa, waiting for a chimp or gorilla to do something astonishing that no one else has ever observed, have learned the hard way that instead many primates will spend endless hours doing nothing but sitting and grooming one another in what looks like an almost meditative trance. The primary purpose of grooming is to get rid of parasites that are in hard to reach places, but it also seems to reduce tension within groups, lowering levels of stress hormones in the recipient, and increasing the circulation of pleasurable chemicals, such as endorphins, in both parties. In some cases, strict social rules of reciprocity seem to govern these grooming rituals. After all, grooming someone takes time, and in a competitive biological marketplace like nature, rather than time being money, it's the currency of survival. It must be doled out with great care. Engaging in any activity that you don't get paid back for sufficiently is risky business, and primates are able to keep close track of their accounts. When Gabriele Schino looked at social grooming in thirty-six studies done in primates, she found individuals were keeping close track of who groomed them, and dispensing grooming as a function of that. Indeed, sometimes they even paid back being groomed in a different currency, such as help finding food or water. You need to know who you can trust in the world of grooming, and animals are acutely aware of their social milieu when they jump into this business.[8]

Other work has revealed that some primates follow social rules governing the formation of coalitions and alliances to get what they want. Baboons have developed a "buddy" system, according to which individuals keep track of who can be trusted and who can't.[9] During mating time, male baboons low on the dominance totem pole will often solicit the aid of another male to gain access to a receptive female that is being guarded by a more dominant male. Craig Packer observed that one baboon will often recruit another to join him in threatening an opponent by repeatedly looking over at the recruit while he continues to make threatening gestures toward the opponent. Sometimes this works, and when it does, the male who did the recruiting is rewarded by being able to mate with the opponent's female mate. The male who joined this alliance gets something in return; those who help are much more likely to get help in return with their own such challenges.[10]

Social cognition in the animal world can also involve deception. In vervet monkeys, when a predator is spotted, individuals emit alarm calls to warn others and some vervets have found a way to use these calls to fool the others in the group and save their own skins. When vervet troops meet along their borders, aggression sometimes breaks out between members of these groups. In 264 intergroup interactions that Dorothy Cheney and Robert Seyfarth recorded, false alarm calls—calls when there is no real danger—were sometimes made by low-ranking males. They seemed to be diverting attention to a mythical predator so that the group focused on that threat rather than on engaging in conflict among themselves, in which these low-ranking males would likely take the worst beating.[11]

Animals clearly understand more about their social milieu than researchers originally thought. Brian Hare had contributed important findings about animal social cognition through his studies on dogs and primates. Research had shown that on one classic social intelligence test—what is known as the object choice test—chimps came up short versus dogs, who performed brilliantly on it.[12] Researchers had found that if they placed two opaque containers on a table, and then, unknown to the chimp, put food under one, it was very difficult to provide a chimp with a visual cue that it could use to figure out where the food was. You could point at the correct container, stare at it, touch it, or even place a marker like a wooden block on it, and chimps just don't get it: they are no more likely to choose the container with food than the one without it. Dogs, on the other hand, are virtual geniuses at this sort of object choice task, and are able to cue in on what chimps seem oblivious to.[13]

Hare had conducted his own studies comparing the ability of chimps and dogs and confirmed just how much smarter dogs were at this task. Then he asked himself: *Why are dogs so good at this?* Maybe it was because dogs spend their whole lives with humans, and learn how to do this sort of thing. Or it could be that all canids—dogs, wolves, and so on—were just good at object choice tests, and that it had nothing to do with "dogginess" per se. The only way to know was to design an experiment, so Brian tested both wolves and dogs on this task. The dogs shined as always, and the wolves seemed clueless as to what was going on.[14] Not all canids could do this. He also tested dog pups of different ages. They all did just fine on the object choice test. He tested dogs who had lots of interactions with humans versus few interactions. They all did fine as well. So, Hare realized, it wasn't the amount of time with humans that made dogs so good at the task.

The obvious conclusion was that dogs seemed to have an innate talent for it. That answers the question at one level, but not another. Why, Brian wondered, do dogs have this innate ability to solve hard social cognition tasks whereas chimps don't? The answer, he surmised, likely had something to do with the fact that dogs had been domesticated. "It is likely," Hare wrote in his 2002 *Science* paper, "that individual dogs that were able to use social cues more flexibly than could their last common wolf ancestor . . . were at a selective advantage."[15] During the process of domestication, dogs that were smart enough to pick up on social cues emitted by their humans would get more food because they could do the things that humans wanted them to do, so humans might toss them more scraps as a reward. They might also be able to pick up on cues humans didn't necessarily want them to pick up on, and occasionally scarf some food not meant for them.

It made perfect sense. The skill in dogs was a beautiful adaptation to their new life situation, selected for by their new human masters. He'd come up with a tidy and beautiful explanation for an important question: just the sort of thing a young scientist dreams of.[16]

His mentor, Wrangham, thought otherwise about Hare's findings. Yes, he told Brian, picking up the skill must have something to do with domestication, but was his adaptationist tale—that animals that were socially smarter were selected by humans—the only possible explanation? Was it necessarily the case that the amazing ability of dogs to pick up on human social cues had been favored by selection? Wrangham thought not. He proposed an alternative hypothesis. Maybe, just maybe, this ability was just an accidental byproduct of domestication.[17] Picking up on human social cues hadn't been selected for, he proposed, it just came along for the ride with other traits that had been selected. Hare decided to take the challenge of testing their competing ideas, and they placed a little wager on who was right.

There was really only one place where Hare could do this test, and that was at the fox farm in Akademgorodok. It was the only place where animals had been domesticated from scratch, and where researchers know *exactly* what sort of selection pressures had been in place, and that selection for social intelligence, per se, had *not* been applied. If Brian was right, both the domesticated foxes and the control foxes should fare poorly on the social intelligence test, because the fox team had never selected foxes based on their social intelligence per se. If Richard was right, and social intelligence was indeed a byproduct of domestication, then the domesticated foxes should show social intelligence on par with dogs, but the control foxes should not. When he contacted Lyudmila, through one of her colleagues, to ask whether she would approve of his conducting the study, she said she would love for him to do so. After scraping together about $10,000 in funding from the Explorers' Club, Hare was off to Akademgorodok. Lyudmila and the Institute research staff and fox farm workers gave him a warm reception, and he was thrilled at how quickly he was accepted into their close-knit circle. He even enjoyed the common mispronunciation of his name among the researchers, as "Brain."

When Hare saw the tame foxes uncontrollably wagging their tails at him, he immediately fell in love with them, like everyone does. Getting down to the task at hand, he decided that he should expand on the object choice test done on dogs and wolves.[18] The test with the foxes would be done using two different experimental set-ups. In the first, which was very similar to the test he had given to the dogs and wolves, he would hide food under one of two cups placed on a table that was about four feet in front of a fox.[19] One of the researchers working with him on the study would point and gaze at the cup that had the food under it, and then which cup, if either, the fox preferred would be recorded. The second type of test would not involve food. Instead, two identical toys, that the foxes knew and loved, would be placed on the right and left end of a table placed in front of a fox pup in its home pen.

Hare had the protocols all sketched out and ready to go, when a series of unexpected problems emerged. For one thing, he needed a table to place the cups and toys on, which didn't strike him as a problem until he got a taste of some of the relics of the managed economy that had been a hallmark of life in the USSR. When he asked for a table, he was told that one would be made for him by the shop at the Institute. He was to be provided not with any shoddy contraption, but with a marvel of Russian engineering that even Belyaev would be proud of. The job order went in, and two weeks later, the table arrived. "It was the most beautiful thing you've ever seen," Hare recalls fondly. "I dubbed it 'Sputnik,' which everyone thought was hilarious."[20]

The second problem to be solved before the experiments could begin was a bit trickier. For the test to be fair, the fox had to be standing in the middle of its pen at the start, not on the right or left side. But how could he make sure they were? Some on the fox team suggested he train them to stay in the middle, which they assured him would be possible, but he didn't have time for that, and what's more he wanted to avoid the training procedure as an experimental confound. Instead he thought that if he put a wood plank on the floor of each pen, in the middle, the foxes might prefer to sit or stand on it rather than the chicken wire of the pen's floor. Once again the Institute provided for his needs, and after boards had been placed in the pen of each fox to be tested, when Hare went to the farm the next day, he remembers quite vividly that every single fox was lying on the boards in the middle.

He tested seventy-five fox pups, each one many times.[21] The results were crystal clear. When tame pups were compared to dog pups, they were just as smart as the dogs. And when the tame pups were compared to control fox pups, they were smarter—much smarter—both at finding the hidden food in the pointing and gazing task, and at touching the same toy that Brian or his assistant had touched.[22]

The results were completely in line with Wrangham's hypothesis. The control foxes were clueless on the social cognition tasks, which the domesticated foxes aced, performing even a little better than dogs. Social intelligence, somehow or another, just came along for the ride in their domestication.

"Richard was right," Hare admits, "and I was wrong . . . it totally rocked my world."[23] Suddenly, he saw the evolution of intelligence, and the process of domestication also, very differently. He had thought that early humans intentionally breeding dogs to be smarter had led to dogs' social intelligence. But if the trait could emerge, instead, from selection for tameness, then that was evidence in support of the view that the domestication of the wolf might have started without breeding for social intelligence being involved. Hare now believed that selection acting on tameness could have brought wolves onto the path to domestication, because those that were inherently a little tamer, and began hanging around human groups, would have had the survival advantage of more plentiful food. Wolves might have started the process of domestication themselves, just as Dmitri Belyaev had conjectured, and had argued about human domestication too. This change of understanding was what led Hare to collaborate later with Richard Wrangham on their study of bonobo self-domestication.

Lyudmila knew Dmitri would have been delighted with Brian's finding: the results were entirely in keeping with the theory of destabilizing selection. Shake up the fox genome by placing foxes in a new world where calm behavior toward humans is the ultimate currency, and you'll get lots of other changes—mottled fur, curly, wagging tails, and better social cognition as well.

Hare's work on social cognition inspired one of the members of the fox team to test how well the tame foxes could *learn* to perform many of the tasks dogs have been trained to do. Irena Muchamedshina, who had long been immersed in training her own pet dogs, joined the fox team as a nineteen-year-old undergraduate student at Novosibirsk State University. After she'd been working on the farm for a time, she recalls, "I had seen these foxes daily wiggling their tails and jumping to get a tiniest bit of human attention, and got really curious about the possibility of working with them the same way as I did with dogs."[24] She got Lyudmila's approval to raise one of the pups from the elite tamed line, named Wilj'a, in her little flat from the time that Wilj'a was just six weeks old, so she could start the training young. She also worked every day with another tame pup, Anjuta, at the farm. Each day for three weeks she spent fifteen minutes rewarding the animals with tasty treats for responding correctly to commands such as "sit," "lie down," and "stand up." Both pups learned to recognize the commands quickly and they performed the tasks with the discipline of show dogs. This gave Lyudmila more hope that in time she would be able to convince people to take pups of the elite foxes into their homes. If they could be taught to perform at command so easily, they could almost surely be taught to be consummate house pets.

ANIMAL BEHAVIOR RESEARCHERS ALSO made great headway in the 1980s and '90s in understanding animal communication. Lyudmila knew of this work and was hopeful—up to this point, she had not been able to study the new "ha ha" vocalizations the tame foxes were making, but maybe now that would change.

The bar was long set high for claims about animal communication, especially communication between humans and nonhumans, because of a horse named Clever Hans. At the turn of the twentieth century, William von Osten became something of a celebrity because of the prodigious abilities he claimed for his horse, Clever Hans. Von Osten asserted that Hans could solve mathematical puzzles, identify different pieces of music, and answer questions regarding European history. Of course, Hans didn't talk; he just used his hoof to tap out the solution to a math problem, or shook his head up and down or sideways, "yes" or "no," to questions. The Prussian Academy of Science got wind of von Osten's claims and decided to put Hans to the test in a controlled environment. Hans did give correct answers to the questions thrown at him, but only when someone in the room knew the right answer. If two people each gave Hans part of a question, but each was ignorant of what the other told Hans, Hans did no better than one would expect by pure chance. Hans was indeed clever, just not in the way people thought. He could pick up on very subtle body cues and facial cues that investigators in the room unconsciously emitted when they were giving Hans the correct and incorrect answers to choose from. Animal behaviorists made sure they would not make that mistake.

In the new wave of work, rigorous studies proved that many animals communicate in elaborate ways. Vervet monkeys, again, provide a nice example. Life in the Amboseli National Park in southern Kenya can be dangerous for vervets. Leopards lurk in the bushes; crowned eagles, who can swoop down and carry off a monkey in their talons, scour the landscape for them; and deadly snakes are also about. Fortunately for the vervets, they are able to communicate about such threats with one another. They do this in a remarkable fashion. Vervets send specific alarm calls to one another about different types of danger. If an eagle is spotted, vervets emit a call that sounds to us something like a cough. Vervets who hear this look into the air or hide in the bushes, where they are safe from threats from the sky. If a leopard is seen, but not otherwise, they make a sound more like a bark, and the monkeys respond by climbing trees, where leopards have trouble following them. When a python or cobra is sighted hiding in the tall grass, vervets give off a "chutter" call, at which other vervets stand and scan the grass around them for snakes. For each specific signal a vervet gives there is a specific, adaptive, response by those who receive the signal.[25]

Animal communication wasn't an area Lyudmila had any expertise in, though she found it very interesting. She and her team had long made note of a range of new vocalizations the foxes had begun making, starting with the whining and whimpering of the elite fox pups for human attention and including different bark-like sounds. There was also Coco's chuckling "co, co, co" sound, and the strange "haaaaaw, haaaaaaw, haw, haw, haw"—ha ha—sound that reminded Lyudmila of laughter. No researchers at the Institute had any knowledge of how to study these vocalizations, so Lyudmila had never attempted a study. Then, in 2005, she got a call from someone who wanted to do so.

At the time, twenty-year-old undergraduate Svetlana Gogoleva, who goes by Sveta, was working in the lab of Ilya Volodin, a young professor who worked on animal communication at Lyudmila's alma mater, Moscow State University.[26] Sveta read about the fox experiment and thought it presented a unique opportunity to study how domestication affected the evolution of animal communication abilities. Volodin liked the idea and Sveta and he contacted Lyudmila proposing to record all of the foxes' vocalizations so that she could compare those of the elites, the controls, and the aggressive foxes. As ever, Lyudmila was delighted to welcome Sveta to the fox team.

The first step, Lyudmila told her, should be to have a member of the fox team make a few preliminary recordings of the vocalizations of the elite, control, and aggressive foxes. She'd send the tapes to them at Moscow State and see what Sveta and her professor thought of them. When Sveta and Volodin listened to these tapes, they were fascinated. They had never heard sounds like those the tame foxes were making. "As soon as we analyzed the first recordings," Sveta remembers, "we decided that I just have to go to the farm and begin the work with these unique animals." She began her work at the fox farm in the summer of 2005. "I was a bit nervous," she recalls. After all, she hadn't even completed her undergraduate degree yet. But her anxiety immediately melted away when she met Lyudmila. "At the first sight," she says, "Lyudmila made an impression of a very good and sympathetic person." She invited Sveta to her office and poured them each a cup of tea, and then told Sveta all about Belyaev and the history of the experiment. "Lyudmila was very friendly and often smiled while talking with me," says Sveta, "and because of her smile and soft tone, I soon felt at ease."[27]

Though working with the aggressive foxes was stressful, Sveta loved working with the tame foxes, and she befriended one in particular, named Kefedra. She fondly recalls how when she first went to record Kefreda, the intensely affectionate fox "fell on her side while uttering a long mixed series of cackles and pants," and how when Sveta petted her, Kefedra "tried to push her muzzle into my sleeve and she licked my fingers."

Sveta started her research by cataloguing the different sounds made by the tame, aggressive, and control foxes.[28] "Usually, I began working after the morning feeding of foxes, about 10:00–10:30," she tells it. "I had the list of names and I could freely choose which to test." Right from the start, it was clear that the aggressive foxes were generally louder than the other animals. But, she wasn't especially interested in volume: she wanted to distinguish between the natures of the sounds, and determine whether there were differences among the fox groups. To find out, she tested twenty-five females from each of the tame, control, and aggressive groups.

In each trial, in a well-rehearsed, precise, methodical manner, and armed with a Marantz PM-222 tape recorder, Sveta approached a fox in its home pen. She would stand two to three feet in front of the pen, and if the fox started making sounds, she recorded them for about five minutes. Across the seventy-five females she tested she recorded 12,964 calls, and all of them fell into one of eight categories. Four types of sounds were made by foxes in all groups—tame, control, and aggressive—but of the other four sounds, two were made *only* by the elite foxes, and two *only* by either the aggressive or control foxes.

The two sounds made only by the aggressive foxes and some of the control foxes were vocalizations that sounded (to humans) like a snort and a cough. The vocalizations made *only* by the elite foxes were the cackles and pants she had heard Kefedra make, which were combined in a rapid-fire rhythm of cackle, pant, cackle, pant that produced the strange haaaaw, haaaaaw, haw, haw, haw sound that Lyudmila knew well.

To dig further into her findings, Sveta conducted a detailed analysis of the nature of the cackles and pants—the "ha ha" sound. Analyzing the acoustic microdynamics of the sounds, factoring in things like duration, amplitude, and frequency, she found that, indeed, the combination of the sounds mimicked the sound of human laughter very closely. Closer than any other nonhuman vocalization. When she placed a spectrogram—a visual representation of sound—of the cackles and pants up against a spectrogram of human laughs, she was hard pressed to tell the difference. Lyudmila had been exactly right. The similarity was astonishing. Almost eerie.

The spectrogram analysis led Sveta and Lyudmila to the radically fascinating hypothesis that the tame foxes make the "ha ha" sound in order to attract human attention and prolong interaction with people. Somehow, they propose, the tame foxes have become adept at pleasing us by the sound of our own laughter.[29] How, they don't know, but a more pleasant way for one species to bond with another is hard to imagine.

10

The Commotion in the Genes

For Lyudmila and Dmitri, the fox experiment was, at its heart, about discovering how the genetics of domestication worked. The experiment branched out to encompass many other areas of research, but this had been the core goal from the start. With Anna Kukekova—she who had rushed to the fox farm and collected blood samples in a frenzy of Lyudmila-induced fox domestication science—on board, Lyudmila was finally able to begin probing into the details of the foxes' genome and hoped that analysis would provide further insight into the process of domestication.

The first thing Anna and Lyudmila had to do was to map the fox genome, which was painstaking work. To construct a complete gene sequence would be quite expensive and time consuming, and Anna decided to explore a faster method for creating a somewhat less detailed genomic map. Work on creating a complete sequence of the dog genome was well underway, and Anna wanted to see whether she could make use of the tools developed for analyzing the dog genome, tools called genetic markers,[1] which are stretches of DNA that help to locate, identify, and analyze genes. Because dogs and foxes are evolutionarily close relatives, Anna thought the genomes of the fox and dog might be similar enough that the dog genetic markers would work, but this was by no means a sure thing, given that the ancestors of dogs and those of foxes had separated on the order of 10 million years ago. The genetic makeup of the two had also been found to differ significantly in the number of chromosomes in their genomes. Most breeds of dog have 39 pairs of chromosomes while the silver fox has 17 pairs. Thankfully, in the tedious process of testing 700 genetic markers that had been used in the study of the dog genome, Anna discovered that about 400 of them worked with the fox chromosomes. That was enough firepower to start mapping the fox genome.[2]

When Lyudmila got this news in the fall of 2003, she had recently celebrated her 70th birthday, and the confirmation that the genomic analysis of her foxes could proceed meant a great deal to her. What a long way they all had come, Dmitri, she, and the foxes. When she had first traveled to Novosibirsk to work with Dmitri, they were operating under the shadow of Lysenko, hiding the true nature of their work. Forty-five years later, here she was, hiding nothing. Not only that, but she was collaborating with a Russian—not a Soviet—scientist, who was free to take a job at one of the premier research centers in the US, the Soviet Union's nemesis in the Cold War. And they were making use of tools so sophisticated that researchers could now not only discern the fine line between individual genes, but they could even clone them. If only Dmitri were alive to share in this leg of the journey, she thought.

Using bits of DNA from 286 animals at the fox farm, Anna, Lyudmila and their colleagues meticulously constructed a genomic map of the foxes, which though not comprehensive, covered sections of all sixteen non-sex chromosomes and parts of the female X chromosome, too. Until they had more markers, they wouldn't be able to fill in the rest. They mapped out the relative location of 320 genes in total.[3] While this is a tiny chunk of a typical mammalian genome it was a big step. Now they could begin with the difficult job of identifying which of the genes they had mapped might be linked to the changes involved in domestication, and ultimately begin understanding how on earth it was possible that bits of DNA that once coded for a wild animal could be tweaked to produce a human-loving, domesticated creature. This work would take a good deal more time, and money.[4] Fortunately, their initial results in simply mapping part of the fox genome were promising enough that they were able to secure funds from the National Institutes of Health, who could see some medical implications in understanding the genetic basis of calm, prosocial behavior, as well as that of aggressive, antisocial behavior in the aggressive line of foxes.[5]

While the genomic analysis progressed, Anna reached out to another specialist, Gordon Lark, a professor of biology at the University of Utah. She thought Lark could help her and Lyudmila follow up on Lyudmila's earlier work measuring the differences in the anatomy of the tame and control foxes, which had shown that the snouts of the tame adult foxes were shorter and rounder than those of the control foxes, more like those of fox pups, and also of dogs. Anna knew that Lark and his team had measured the length and width of bones in the body and skull of dogs, and she thought he might agree to help them compare the anatomy of dogs with that of the tame foxes.

Lark's team had discovered that in some dog breeds, animals that had short limbs and short snouts also had wide limbs and wide, rounded snouts—these animals leaned toward a rounded, low to the ground, bulldoggish appearance. Dogs with long graceful limbs and long snouts had relatively narrow snouts, looking more greyhoundish than bull-doggish. Genetic analysis performed by Lark's team suggested that the relationship between the length and width of their bones was controlled by a small number of genes that affect skeletal growth.[6]

Anna asked Gordon if he'd be interested in doing a similar study with the silver foxes at the farm. He said he'd love to. But for that, the fox team would need an X-ray device, and Lyudmila didn't have the funds to buy one. So Lark arranged for a transfer of $25,000 to the Institute of Cytology and Genetics for the purchase. Lyudmila oversaw the project on the Russian end, and she put her colleague and friend, Anastasia Kharlamova, whom Lark likes to call "Lyudmila's lieutenant," in charge of everyday operations. Anastasia began taking X-rays of the bodies and skulls of tame, aggressive, and control foxes, and one of Lark's colleagues set up a website where the X-ray images could be posted, so that the team in Utah could do the analysis of bone width and length, which required their expertise.

This was Lark's initiation into the intensity and efficiency with which Lyudmila's team worked. He recalls, "The volume of data that began to pour in was amazing. It was as if the fox team had fifty hours in a day, somehow." The hard work paid off. Lark's team determined that the same relationship between bone width and length they had found in dogs—short limbs and short snouts paired with wide limbs and wide, rounded snouts—had developed in the foxes.

Lark and Lyudmila proposed an intriguing idea about why these changes had emerged in the foxes. For foxes in the wild, as pups mature and wean off nursing, the body and face shape change in a manner that provides them with the best chance of survival. When they are pups, their faces are relatively round and their legs are chunky. But, as they mature to adulthood, longer more graceful legs provide more speed for chasing prey and evading predators, and longer, more pointed snouts facilitate probing into the nooks and crannies of thick grasses and undergrowth while foraging for food. In wild foxes, this leads to a change in body shape during development, producing the classic anatomy of adult foxes; but on the farm, the foxes never have to forage, hunt, or evade predators, and selection favors juvenile-like traits, hence the more rounded face and chunky body form continue into adulthood in the tame foxes.[7]

WHILE LYUDMILA AND LARK WERE WORKING on this study of the anatomy of the tame foxes, Anna, Lyudmila, and their colleagues proceeded with the next stage of the DNA analysis, which was designed to try to link the genomic work on the foxes with their behavior. DNA samples were taken from 685 tame and aggressive foxes and all these animals were videotaped interacting with a researcher at the fox farm. A meticulous, arguably obsessive, analysis of 98 behaviors was conducted, noting characteristics such as "tame sounds," "aggressive sounds," "tame ears," "aggressive pinned-back ears," "observer can touch fox," "fox comes to sniff observer's hand," "fox rolls onto its side," and "fox invites observer to touch its belly," to name just a few. This project, which came to fruition in 2011, was a tremendous undertaking, but fortunately, the results made all of the work worthwhile.

They discovered that the genes associated with many of the changes to the unique behavioral and morphological characteristics of the tame foxes could be mapped onto a specific region of fox chromosome number 12. On this region, the elite and aggressive foxes had different sets of genes, and Lyudmila, Anna, and their team hypothesized that these genes were likely involved with the changes that distinguished the tame foxes from all others.[8]

Just a year earlier, in 2010, a much heralded paper on dog domestication, published in the prestigious journal *Nature,* had announced that many of the genetic changes that led to the evolution of dogs from wolves could be traced to genes on just a few chromosomes. Now, Anna and Lyudmila could see whether the genetic changes on fox chromosome 12 that distinguished tame foxes from wild foxes were similar to the genetic changes involved in the domestication of dogs. They hoped that they would find considerable similarity in the two sets of genes, and they did. Many of the genes on chromosome 12 of the foxes that were involved in their domestication were also found on the corresponding dog chromosomes involved in their domestication. It was almost too good to be true.

Fifty-nine years after Dmitri took his long train ride to Estonia to meet with Nina Sorkina at the Kohila fox farm to start breeding the first tamer foxes, and fifty-three years after Lyudmila had joined him in the quest, they knew where at least some of the genes associated with fox domestication were located. Next they would conduct experiments to probe into the specific function of each of the genes and into whether the expression of these genes had been altered to bring about the characteristics of domestication, as Dmitri had suggested from the start, before people even had a lexicon with such terms. By 2011, technology was available to begin doing this.

"Next generation sequencing technology" sped up the rate at which DNA sequences could be read, making it possible for millions, sometimes billions, of small bits of DNA to be read by computer analysis rather than by the human eye. Analyzing the effects of genes and how they are being expressed is still an enormously complex process, because genes generally code for different effects in different cells of the body. Each cell in an animal's body, except sperm and eggs, has the same set of genes housed in the pairs of its chromosomes. But different genes are turned on or off in skin cells, say, versus blood cells or brain cells, and some genes that are turned on in more than one type of cell code for the production of different proteins in one cell type versus another. Analyzing the full story of the expression of any given gene in one animal versus another therefore involves comparing the amount of different proteins the gene codes for in all the different types of cells in the body. Researchers typically begin by focusing on a particular type of cell in a particular part of the body. So the first question that Anna and Lyudmila had to tackle was what type of cell they would examine. They decided to begin by studying the expression of the set of genes in the foxes' brain tissue because the brain is the master controller of behavior, and the changes in the foxes had begun with the selection for tameness. The prefrontal cortex had been identified as especially important in controlling behavior, so that is where they drew cells from.[9]

They were able to identify 13,624 genes, and in a complex analysis of the amount of proteins being produced by those genes in the tame foxes versus the aggressive foxes, they discovered that in 335 of these genes—or about 3%—there were dramatic differences in the protein production levels. For example, the HTR2C gene, which is important in the production of serotonin and dopamine, had higher levels of expression in the tame foxes. What was especially intriguing was that with some of the 335 genes—280 of them—expression was higher in the tame foxes than in the aggressive foxes, while in the rest of them, expression was lower in the tame foxes than in the aggressive foxes. So the change to tamer behavior appeared to involve no simple process. What's more, there were complex *interactions* between these genes as well. So complex is the story of the expression of the full set of these genes that it will be the subject of investigation for years to come.

Today, Lyudmila and Anna are still engaged in the delicate and time-intensive process of identifying the specific functions of these 335 genes. They have determined that some are involved in hormone production, others in the development of the blood system, susceptibility to disease, fur and skin development, and the production of vitamins and minerals. The effects on hormone production were expected, because they had discovered so many critical hormonal changes in the tame foxes. How the other effects are related to the behavior of the elite foxes remains a mystery. As more pieces of this complex puzzle are put in place, a clearer picture of the destabilization of the silver fox genome will emerge, and with it a much more refined understanding of the process of wolf and fox domestication.[10]

IN LAUNCHING THE FOX EXPERIMENT Dmitri had theorized that the same fundamental process of selection for tameness was involved in all animal domestications. In the cases of the domestication of the wolf and the fox, he was right that many of the same changes to their genomes and the expression of their genes were likely involved. But how much do these results explain about the process of domestication in other species? Are the same genes and changes in their expression involved?

A recent analysis by Frank Albert and a team of geneticists that included Lyudmila compared the genes involved in the domestication of three species—dogs, pigs, and rabbits—and the expression levels of those genes in the domesticated animals versus the ancestral animals of each—wolves, wild boars, and wild rabbits, respectively. The researchers found little evidence that the exact same set of genes and the same changes in their expression were involved. They did find that two genes associated with brain development might be commonly involved in all three cases of domestication, and further work is underway on that tantalizing finding.[11]

While for now, the process of the domestication of other species, including that of us humans, remains shrouded in mystery, in principle at least, we should be able to solve the riddle for them all with time. The better the techniques of genetic analysis become and the more that archeology, anthropology, and genetics shed light on the history of the domestication of other species, the more we will understand about how similar the process might have been across species, and whether Dmitri Belyaev was right that selection for tameness and destabilizing selection was behind all cases.

Though the specific genes involved in different species may differ, there are hints that Belyaev was right that the process is similar across species in key ways. Work on domestication genes in many species shows that domestication involves just the sort of complicated set of genetic changes that Belyaev described in his theory of destabilizing selection. For example, work on the domestication of rabbits in Southern France, found that "at least some of the selection occurred on genetic variation that already existed in the population, rather than on new mutations," just as Dmitri predicted.[12] And much of the work being done on domestication shows that, as with the foxes, the expression of genes, not just their presence or absence, is key to domestication.

Also providing some support for Belyaev's destabilizing selection theory is a promising new theory proposed by Adam Wilkins, Richard Wrangham, and Tecumseh Fitch on why selection for tameness would lead to a cascade of other new traits. They propose that changes to a type of stem cell, called a neural crest cell, may help explain many of the traits that domesticated species share. Very early on in vertebrate embryonic development, these cells move along what is known as the neural crest—a concentration of neurons in the middle of the developing embryo—and migrate to different parts of the body, such as the forebrain, skin, jaws, teeth, larynx, ears, and cartilage. Wilkins and his colleagues hypothesize that selection for tameness may also select for a small reduction in the number of neural crest cells, and that "Most of the modified traits, both morphological and physiological [associated with domestication], can be readily explained as direct consequences of such deficiencies [in neural crest cells], while other traits are explicable as indirect consequences."[13] Exactly how this might occur is unclear, but if correct this might help explain how tameness is linked to the whole suite of traits we see in domesticated species—mottled coloring, floppy ears, shorter snouts, changes in reproduction, a curly tail and so on. It is an intriguing hypothesis and requires further investigation.

EVENTUALLY, THE FOX EXPERIMENT will produce many more exciting discoveries. The experiment has been going on for almost sixty years now, which is eons for a biology experiment. But from an evolutionary perspective, sixty years is only the blink of an eye. What would happen if the experiment ran 100 generations? Or 500 generations? Are there limits to how tame and how symbiotically habituated to life with humans the foxes would become? How much more dog-like in appearance would they get? How smart might they grow to be? Would they develop into staunch guardians, as Pushinka's bark in the dark to alert and defend Lyudmila might suggest? And perhaps, just perhaps, as Dmitri Belyaev hoped, the work with the foxes will ultimately help to explain how, deep in the chromosomes, a stirring occurred that set the common ancestors of all of the other domesticates on the road to tameness, including the ancestors of humans.

One thing about the domestication of the foxes that has already been definitively determined is that they have become a new line of animal that we humans can take into our lives and love. This, in fact, is Lyudmila's great hope for her foxes, who have become, in her words, such "dainty, fluffy, charming rogues."

In 2010, Lyudmila started to seriously explore whether people might want to purchase tame foxes as pets, and a number of foxes have been adopted and are living happily with families in Russia, Western Europe, and North America. The owners write to Lyudmila sometimes to update her about how they and the foxes are doing, which delights her. She likes to pull these letters out on occasion and read them again, smiling about the escapades the owners recount and their affection for the foxes.

One American couple who adopted two foxes, named Yuri and Scarlet, wrote recently that the pair "play well together and are both very social. They both enjoy getting out and seeing everything possible!"[14] Another letter arrived recently about a close call experienced by a fox named Arsi : "Arsi . . . had a little accident about a week ago. He stopped eating for a couple of days and threw up a couple of times. I took him in [to the vet] for blood work and an x-ray. [The vet] removed a piece of a rubber toy shaped like a V that came off a ball I had bought him. It's like looking after a child because they do put everything in their mouths!"

All the letters are special to Lyudmila, but one stands out. "Hi Lyudmila, I am very happy," the letter begins. The owner adopted a fox named Adis and he reports that "Adis is wonderful . . . when I come home from work Adis wags his tail and likes to kiss me."[15] *Kiss me,* she thinks every time she reads the letter, *how wonderful. How Dmitri would have savored that.*

Having celebrated her 83rd birthday in 2016, Lyudmila is still working with the foxes. The wise words uttered by the fox in Saint-Exupéry's *The Little Prince* that "you become responsible forever for what you tame," are Lyudmila's constant companions. Her dream is to establish a secure and loving future for the foxes. "I hope that it is possible to register them as a new pet species," Lyudmila says. "One day I will be gone, but I want my foxes to live forever." She knows that convincing more people to take the foxes into their homes won't be easy. But easy doesn't matter to Lyudmila. Easy never has mattered. Possible is what matters.

Acknowledgments

First and foremost, we thank Dmitri Belyaev for his brilliant insights, and for launching, more than six decades ago, what was an audacious experiment to domesticate the silver fox. Dmitri has been gone for more than thirty years, yet hardly a day passes when the fox research team in Siberia does not think of this wonderful man and wish he were still there guiding them. He had few regrets when he left, save one; that he had not written his popular book *Man Is Making a New Friend*, which is truly the heart of the story in this book. One look into the eyes of the tamed foxes, one lick from their loving tongues and wag of their bushy tails, and no one could have any doubt—we humans have indeed made a lovable, and deeply loyal, new friend.

We hardly know where to begin when it comes to thanking all the people who helped us put this book together. We are deeply indebted to Tamara Kuzhutova, Lyudmila's dear friend and colleague, who has participated in the fox experiment from its earliest days. We extend sincere thanks to Ekaterina Omelchenko, who, over many years, has mined the experimental data and created an electronic database. We thank Pavel Borodin, Anatoly Ruvinsky, Michael

(Misha) Belyaev, Nicholai Belyaev, Svetlana Argutinskaya, and Arkady Markel for all they have done over the years, both as Lyudmila's collaborators and as her friends. Over the course of this experiment hundreds of researchers have been involved in one way or another, and while we can't thank them all, we would be remiss not to acknowledge the amazing work of Irena Plysnina, Irena Oskina, Lyudmila Prasolova, Larisa Vasilyeva, Larisa Kolesnikova, Anastasia Kharlamova, Rimma Gulevich, Juriy Gerbek, Lyudmila Kondrina, Klaudia Sidorova, Vasily Evaikin (chief of the fox farm), Ekaterina Budashkinah, Natasha Vasilevskaya, Irena Muchamedshina, Darja Shepeleva, Anastasia Vladimirova, Svetlana Shikhevich, Irena Pivovarova, Tatjana Semenova, and Vera Chaustova (long-term veterinarian on the fox project). We are also deeply indebted to Venya and Galya Esakovi for their love, care, and kindness toward Coco the fox, who lived with them at their home for much of her life.

While it might seem a tad odd, as coauthors, we would like to thank each other. Lee wishes to thank Lyudmila for her friendship, for allowing him the sheer delight of getting involved in one of the most important science experiments ever undertaken, and for the chance to get to know all the remarkable people involved in this work. Lyudmila wishes to thank Lee for his friendship, for making the journey across time and space to visit her more than once at the fox farm to meet the foxes she cares about so deeply and to talk with the many dear friends and colleagues of Dmitri Belyaev about their memories of him and of all the exciting moments of discovery with the domesticated foxes.

We are indebted to the following individuals for allowing us to interview them for insight into the people, the wonderful foxes, and the path-breaking science we write of: Anatoly Ruvinksy, Pavel Borodin, Michael (Misha) Belyaev, Nicholai Belyaev, Larisa Vasilyeva, Valery Soyfer, Galina Kiseleva, Vladimir Shumny, Larisa Kolesnikova, Natalie Delaunay, Anna Kukekova, Svetlana Gogoleva, Ilya Ruvinsky, Nicholai Kolchanov, L. V. Znak, Oleg Trapezov, Aubrey Manning, John Scandalious, Brian Hare, Gordon Lark, Francisco Ayala, Bert Hölldobler, Marc Bekoff, and Gordon Burghardt.

The day-to-day maintenance of hundreds of foxes year after year for close to six decades is an expensive endeavor. Lyudmila is especially grateful to Vladimir Shumny, who served as director of the Institute of Cytology and Genetics from 1985 to 2007, and Nicholai Kolchanov, who serves in that role today, both of whom provided critical financial aid that allowed the fox work to continue through very rough times.

We thank Susan Rabiner, of the Susan Rabiner Literary Agency, for helping us shape what you have read here. Our editor at the University of Chicago Press, Christie Henry, is as good as they come and has been a delight to work with on this project, and we appreciate the help of Christie's editorial assistant, Gina Wadas, two anonymous readers of the manuscript, and the Editorial Board at the University of Chicago Press. Comments on various chapters of the book were graciously provided by Pavel Borodin, Carl Bergstrom, Henry Bloom, John Shumate, Aaron Dugatkin, Dana Dugatkin, Michael Sims, and most especially, Emily Loose. Dana Dugatkin transcribed interviews and proofread the manuscript an inordinate number of times; we are indebted to her for all her suggestions. We thank Aaron Dugatkin, who accompanied Lee on the journeys to Akademgorodok in Siberia, transcribed interviews, and enjoyed a daily lunch of shashlik with Lee at *Vkusnyy Tsentr*. While we were at Akademgorodok, Vladimir Filonenko acted as our intrepid interpreter, and Egor Dyomin shepherded our team from place to place, gliding us gently along the ice-encrusted roads of Novosibirsk. When not in Siberia, we received unparalleled help in translating between Russian and English from Amal El-Sheikh, culture and language consultant at Alverno College. On occasion, Tom Dumstorf at the University of Louisville also provided help with translation.

Notes

Chapter 1

1. Here Belyaev was influenced by the work of one of his intellectual idols, Nikolai Vavilov, in particular Vavilov's Law of Homologous Series.
2. In Russian, this sort of settlement is called a Poselok (поселок).
3. Summarized in his book, N. I. Vavilov, *Five Continents* (Rome: IPGRI, English translation, 1997).
4. Vavilov Research Institute: http://vir.nw.ru/history/vavilov.htm#expeditions.
5. S. C. Harland, "Nicolai Ivanovitch Vavilov, 1885–1942," *Obituary Notices of Fellows of the Royal Society* 9 (1954): 259–264.
6. D. Joravsky, *The Lysenko Affair* (Cambridge: Harvard University Press, 1979); V. Soyfer, "The Consequences of Political Dictatorship for Russian Science," *Nature Reviews Genetics* 2 (2001): 723–729; V. Soyfer, *Lysenko and the Tragedy of Soviet Science* (New Brunswick: Rutgers University Press, 1994); V. Soyfer, "New Light on the Lysenko Era," *Nature* 339 (1989): 415–420.
7. From the Agricultural Institute of Kiev.
8. From the Agricultural Institute of Kiev.
9. Vitaly Fyodorovich.
10. Soyfer, *Lysenko and the Tragedy of Soviet Science.*
11. Soyfer, *Lysenko and the Tragedy of Soviet Science,* 11; *Pravda,* October 8, 1929, as cited on in Soyfer, *Lysenko and the Tragedy of Soviet Science.*

12. *Pravda*, February 15, 1935; *Izvestia*, February 15, 1935. As cited in Joravsky, *The Lysenko Affair*, 83, and Soyfer, *Lysenko and the Tragedy of Soviet Science*, 61.
13. Z. Medvedev, *The Rise and Fall of T. D. Lysenko* (New York: Columbia University Press, 1969).
14. Medvedev, *The Rise and Fall of T. D. Lysenko.*
15. P. Pringle, *The Murder of Nikolai Vavilov* (New York: Simon and Schuster, 2008), 5.
16. Founded in 1916, this institute was funded by a private charity and was part of the People's Commissariat of Health; S. G. Inge-Vechtomov and N. P. Bochkov, "An Outstanding Geneticist and Cell-Minded Person: On the Centenary of the Birth of Academician B. L. Astaurov," *Herald of the Russian Academy of Sciences* 74 (2004): 542–547.
17. S. Argutinskaya, "Memories," in *Dmitry Konstantinovich Belyaev*, ed. V. K. Shumny, P. Borodin, and A. Markel (Novosibirsk: Russian Academy of Sciences, 2002), 5–71.
18. Argutinskaya, "Memories." Their son was left to fend for himself, eventually ending up in the care of an aunt.
19. Joravsky, *The Lysenko Affair*, 137.
20. T. Lysenko, "The Situation in the Science of Biology" (address to the All-Union Lenin Academy of Agricultural Sciences, July 31–August 7, 1948). The entire speech, in English, can be found at http://www.marxists.org/reference/archive/lysenko/works/1940s/report.htm.
21. From stenographic notes of 1948 meeting, *O polozhenii v biologicheskoi nauke. Stenograficheskii otchet sessi VASKhNILa 31 iiula-7 avgusta 1948.*
22. Argutinskaya, "Memories."
23. Argutinskaya, "Memories."

Chapter 2

1. K. Roed, O. Flagstad, M. Nieminen, O. Holand, M. Dwyer, N. Rov, and C. Via, "Genetic Analyses Reveal Independent Domestication Origins of Eurasian Reindeer," *Proceedings of the Royal Society of London B* 275 (2008): 1849–1855.
2. Soyfer, *Lysenko and the Tragedy of Soviet Science.*
3. As cited in *Scientific Siberia* (Moscow: Progress, 1970).
4. The head of the committee was M. A. Olshansky.
5. Trofimuk's memoirs of Khrushchev's visits: http://www-sbras.nsc.ru/HBC/2000/n30–31/f7.html.

6. Ekaterina Budashkinah, interview with authors, January 2012.
7. I. Poletaeva and Z. Zorina, "Extrapolation Ability in Animals and Its Possible Links to Exploration, Anxiety, and Novelty Seeking," in *Anticipation: Learning from the Past,*" ed. M. Nadin (Berlin: Springer, 2015), 415–430.

Chapter 3

1. D. Belyaev to M. Lerner, July 15, 1966. From collection of Lerner letters at the American Philosophical Society.
2. P. Josephson, *New Atlantis Revisited: Akademgorodok, The Siberian City of Science* (Princeton: Princeton University Press, 1997).
3. Josephson, *New Atlantis Revisited,* 110.
4. L. Trut, I. Oskina, and A. Kharlamova, "Animal Evolution during Domestication: The Domesticated Fox as a Model," *Bioessays* 31 (2009): 349–360.
5. The term *destabilizing selection* has other meanings as well within the field of evolutionary biology.
6. Tamara Kuzhutova, interview with authors, January 2012.
7. M. Nagasawa et al., "Oxytocin-Gaze Positive Loop and the Coevolution of Human-Dog Bonds," *Science* 348 (2015): 333–336; A. Miklosi et al., "A Simple Reason for a Big Difference: Wolves Do Not Look Back at Humans, but Dogs Do," *Current Biology* 13 (2003): 763–766.
8. B. Hare and V. Woods, "We didn't domesticate dogs, they domesticated us," 2013, http://news.nationalgeographic.com/news/2013/03/130302-dog-domestic-evolution-science-wolf-wolves-human/.
9. C. Darwin, *The Expression of Emotions in Man and Animals,* 2nd ed. (London: J. Murray, 1872).
10. N. Tinbergen, *The Study of Instinct* (Oxford: Clarendon Press, 1951); N. Tinbergen, "The Curious Behavior of the Stickleback," *Scientific American* 187 (1952): 22–26.
11. K. Lorenz, "Vergleichende Bewegungsstudien an Anatiden," *Journal fur Ornithologie* 89 (1941): 194–293; K. Lorenz, *King Solomon's Ring,* trans. Majorie Kerr Wilson (London: Methuen, 1961). Original in German is from 1949.

Chapter 4

1. A. Forel, *The Social World of the Ants as Compared to Man,* vol. 1 (New York: Albert and Charles Boni, 1929), 469.

2. T. Nishida and W. Wallauer, "Leaf-Pile Pulling: An Unusual Play Pattern in Wild Chimpanzees," *American Journal of Primatology* 60 (2003): 167–173.
3. A. Thornton and K. McAuliffe, "Teaching in Wild Meerkats," *Science* 313 (2006): 227–229.
4. B. Heinrich and T. Bugnyar, "Just How Smart Are Ravens?" *Scientific American* 296 (2007): 64–71; B. Heinrich and R. Smokler, "Play in Common Ravens (*Corvus corax*)," in *Animal Play: Evolutionary, Comparative and Ecological Perspectives*, ed. M. Bekoff and J. Byers (Cambridge: Cambridge University Press, 1998), 27–44; B. Heinrich, "Neophilia and Exploration in Juvenile Common Ravens, *Corvus corax*," *Animal Behaviour* 50 (1995): 695–704.
5. L. Trut, "A Long Life of Ideas," in *Dmitry Konstantinovich Belyaev*, 89–93.
6. D. Belyaev, A. Ruvinsky, and L. Trut, "Inherited Activation-Inactivation of the Star Gene in Foxes: Its Bearing on the Problem of Domestication," *Journal of Heredity* 72 (1981): 267–274.
7. Thirty-five percent of the variation observed was due to genetic variation: L. Trut and D. Belyaev, "The Role of Behavior in the Regulation of the Reproductive Function in Some Representatives of the Canidae Family," in *Vie Congres International de Reproduction et Insemination Artificielle* (Paris: Thibault, 1969), 1677–1680; L. Trut, "Early Canid Domestication: The Farm-Fox Experiment," *American Scientist* 87 (1999): 160–169.
8. F. Albert et al., "Phenotypic Differences in Behavior, Physiology and Neurochemistry between Rats Selected for Tameness and for Defensive Aggression towards Humans," *Hormones and Behavior* 53 (2008): 413–421.
9. Svetlana Gogolova, email interview with authors.
10. Natasha Vasilevskaya, interview with authors, January 2012.
11. Aubrey Manning, Skype interview with authors.
12. Aubrey Manning, Skype interview with authors.
13. People such as John Fentress, J. P. Scott, Bert Hölldobler, Patrick Bateson, Klaus Immelman, and Robert Hinde.
14. Bert Hölldobler, Skype interview with authors. Hölldobler attended the 1971 meeting.
15. D. Belyaev, "Domestication: Plant and Animal," in *Encyclopedia Britannica*, vol. 5 (Chicago: Encyclopedia Britannica, 1974): 936–942.
16. R. Levins, "Genetics and Hunger," *Genetics* 78 (1974): 67–76; G. S. Stent, "Dilemma of Science and Morals," *Genetics* 78 (1974): 41–51.
17. *Genetics* 79 (June 1975 supplement): 5.
18. S. Argutinskaya, "D. K. Belyaev, 1917–1985, from the First Steps to Founding the Institute of Cytology and Genetics of Siberian Branch of the Russian Academy of Sciences of USSR (ICGSBRAS)," *Genetika* 33 (1997): 1030–1043.

Chapter 5

1. P. McConnell, *For the Love of a Dog* (New York: Ballantine, 2007).
2. A. Horowitz, "Disambiguating the 'Guilty Look': Salient Prompts to a Familiar Dog Behavior," *Behavioural Processes* 81 (2009): 447–452; C. Darwin, *The Expression of Emotions in Man and Animals* (London: J. Murray, 1872); K. Lorenz, *Man Meets Dog* (Methuen: London, 1954); H. E. Whitely, *Understanding and Training Your Dog or Puppy* (Santa Fe: Sunstone, 2006); D. Cheney and R. Seyfarth, *Baboon Metaphysics: The Evolution of a Social Mind* (Chicago: University of Chicago Press, 2007); F. De Waal, *Good Natured: The Origins of Right and Wrong in Humans and Other Animals* (Cambridge: Harvard University Press, 1997).
3. A. Horowitz, "Disambiguating the 'Guilty Look.'"
4. J. van Lawick-Goodall and H. van Lawick, *In the Shadow of Man* (New York: Houghton-Mifflin, 1971).
5. P. Miller, "Crusading for Chimps and Humans," National Geographic website, December 1995, http://s.ngm.com/1995/12/jane-goodall/goodall-text/1.

Chapter 6

1. A. Miklosi, *Dog Behaviour, Evolution, and Cognition* (Oxford: Oxford University Press, 2014).
2. M. Zeder, "Domestication and Early Agriculture in the Mediterranean Basin: Origins, Diffusion, and Impact," *Proceedings of the National Academy of Sciences* 15 (2008): 11587–11604; "Domestication Timeline," American Museum of Natural History website, http://www.amnh.org/exhibitions/past-exhibitions/horse/domesticating-horses/domestication-timeline.
3. M. Deer, "From the Cave to the Kennel," Wall Street Journal website, October 29, 2011, http://www.wsj.com/articles/SB10001424052970203554104577001843790269560.
4. M. Germonpre et al., "Fossil Dogs and Wolves from Palaeolithic Sites in Belgium, the Ukraine and Russia: Osteometry, Ancient DNA and Stable Isotopes," *Journal of Archaeological Science* 36 (2009): 473–490.
5. E. Axelsson et al., "The Genomic Signature of Dog Domestication Reveals Adaptation to a Starch-Rich Diet," *Nature* 495 (2013): 360–364.
6. R. Bridges, "Neuroendocrine Regulation of Maternal Behavior," *Frontiers in Neuroendocrinology* 36 (2015): 178–196; R. Feldman, "The Adaptive Human Parental Brain: Implications for Children's Social Development," *Trends in Neurosciences* 38 (2015): 387–399; J. Rilling and L. Young, "The Biology of

Mammalian Parenting and Its Effect on Offspring Social Development," *Science* 345 (2014): 771–776.

7. S. Kim et al., "Maternal Oxytocin Response Predicts Mother-to-Infant Gaze," *Brain Research* 1580 (2014):133–142; S. Dickstein et al., "Social Referencing and the Security of Attachment," *Infant Behavior & Development* 7 (1984): 507–516.
8. J. Odendaal and R. Meintjes, "Neurophysiological Correlates of Affiliative Behaviour between Humans and Dogs," *Veterinary Journal* 165 (2003): 296–301; S. Mitsui et al., "Urinary Oxytocin as a Noninvasive Biomarker of Positive Emotion in Dogs," *Hormones and Behavior* 60 (2011): 239–243.
9. M. Nagasawa et al., "Oxytocin-Gaze Positive Loop and the Coevolution of Human-Dog Bonds"; M. Nagasawa et al., "Dog's Gaze at Its Owner Increases Owner's Urinary Oxytocin during Social Interaction," *Hormones and Behavior* 55 (2009): 434–441.
10. The name *serotonin* was not adopted until later.
11. G. Z. Wang et al., "The Genomics of Selection in Dogs and the Parallel Evolution between Dogs and Humans," *Nature Communications* 4 (2013), DOI:10.1038/ncomms2814.
12. Descartes in a letter dated January 29, 1640; see Descartes's *View of the Pineal Gland* in "The Stanford Encyclopedia of Philosophy," http://plato.stanford.edu/entries/pineal-gland/#2.
13. Larissa Kolesnikova, phone interview with authors.
14. Larissa Kolesnikova, phone interview with authors.
15. L. Kolesnikova et al., "Changes in Morphology of Silver Fox Pineal Gland at Domestication," *Zhurnal Obshchei Biologii* 49 (1988): 487–492; L. Kolesnikova et al., "Circadian Dynamics of Biochemical Activity in the Epiphysis of Silver-Black Foxes," *Izvestiya Akademii Nauk Seriya Biologicheskaya* (May-June 1997): 380–384; L. Kolesnikova, "Characteristics of the Circadian Rhythm of Pineal Gland Biosynthetic Activity in Relatively Wild and Domesticated Silver Foxes," *Genetika* 33 (1997): 1144–1148; L. Kolesnikova et al., "The Melatonin Content of the Tissues of Relatively Wild and Domesticated Silver Foxes *Vulpes fulvus*," *Zhurnal Evoliutsionnoĭ Biokhimii i Fiziologii* 29 (1993): 482–496.
16. John Scandalious, phone interview with authors.
17. N. Tsitsin, "Presidential Address: The Present State and Prospects of Genetics," in *XIV International Congress of Genetics*, ed. D. K. Belyaev, vol. 1 (Moscow: MIR Publishers, 1978), 20.
18. Penelope Scandalious's journal, personal communication with authors.
19. M. King and A. Wilson, "Evolution in Two Levels in Humans and Chimpanzees," *Science* 188 (1975): 107–116; when it came to gene expression and mutation, they were referring to changes associated with point mutations.

20. Aubrey Manning, Skype interview with authors.
21. Aubrey Manning, Skype interview with authors.

Chapter 7

1. L. Mech and L. Boitani, eds., *Wolves: Behavior, Ecology, and Conservation* (Chicago: University of Chicago Press, 2007).
2. J. Goodall to W. Schleidt, as cited in "Co-evolution of Humans and Canids," *Evolution and Cognition* 9 (2003): 57–72.
3. L. S. B. Leakey, "A New Fossil from Olduvai," *Nature* 184 (1959): 491–494.
4. A multiregional theory of human evolution that is still championed by some today, with heated debate between them and the dominant "Out of Africa" camp of the research community. The multiregional hypothesis posits that *Homo erectus* left Africa and colonized the Old World a single time, nearly 2 million years ago, then *H. erectus* populations diverged from one another, then over the past 2 million years, these loosely associated populations together evolved into modern humans. The out-of-Africa hypothesis posits that hominins experienced two major waves out of Africa, colonizing first as *Homo erectus* about 2 million years ago, and then as *Homo sapiens* approximately 100,000 years ago. Modern *Homo sapiens* emerged in Africa, and in the second wave of colonization, the pre-modern hominins of Europe and Asia, such as *Homo erectus* and *Homo neanderthalensis*, were replaced by *Homo sapiens*. Modified from C. Bergstrom and L. Dugatkin, *Evolution* (New York: W. W. Norton, 2012).
5. Later revised to 3.2 million years ago.
6. D. K. Belyaev, "On Some Factors in the Evolution of Hominids," *Voprosy Filosofii* 8 (1981): 69–77; D. K. Belyaev, "Genetics, Society and Personality," in *Genetics: New Frontiers: Proceedings of the XV International Congress of Genetics*, ed. V. Chopra (New York: Oxford University Press, 1984), 379–386.
7. But now is dated between 1.5 and 2 million years ago.
8. D. K. Belyaev, "On Some Factors in the Evolution of Hominids."
9. D. K. Belyaev, "Genetics, Society and Personality."
10. The notion of human self-domestication had been mentioned on occasion before Belyaev, but not in a systematic or detailed manner. W. Bagehot, *Physics and Politics or Thoughts on the Application of the Principles of "Natural Selection" and "Inheritance" to Political Society* (London: Kegan Paul, Trench and Trubner, 1905). In addition, subsequently, self-domestication of humans has been used to describe a process that is quite different from what Belyaev was dis-

cussing: P. Wilson, *The Domestication of the Human Species* (New Haven: Yale University Press, 1991).

11. B. Hare, V. Wobber, and R. Wrangham, "The Self-Domestication Hypothesis: Evolution of Bonobo Psychology Is Due to Selection against Aggression," *Animal Behaviour* 83 (2012): 573–585. Related papers include B. Hare et al., "Tolerance Allows Bonobos to Outperform Chimpanzees on a Cooperative Task," *Current Biology* 17 (2007): 619–623; V. Wobber, R. Wrangham, and B. Hare, "Bonobos Exhibit Delayed Development of Social Behavior and Cognition Relative to Chimpanzees," *Current Biology* 20 (2010): 226–230; V. Wobber, R. Wrangham, and B. Hare, "Application of the Heterochrony Framework to the Study of Behavior and Cognition," *Communicative and Integrative Biology* 3 (2010): 337–339; R. Cieri et al., "Craniofacial Feminization, Social Tolerance, and the Origins of Behavioral Modernity," *Current Anthropology* 55 (2014): 419–443.
12. D. Quammen, "The Left Bank Ape," National Geographic website, March 2013, http://ngm.nationalgeographic.com/2013/03/125-bonobos/quammen-text.
13. For a map of what this looks like see: http://mappery.com/map-of/African-Great-Apes-Habitat-Range-Map.
14. J. Rilling et al., "Differences between Chimpanzees and Bonobos in Neural Systems Supporting Social Cognition," *Social Cognitive and Affective Neuroscience* 7 (2012): 369–379.
15. There is also some evidence that changes associated with self-domestication in bonobos are driven by changes in the expression and timing of regulatory genes associated with the stress hormone system, just as Belyaev thought. The exact role of gene regulation across domesticated species is still unclear: F. Albert et al., "A Comparison of Brain Gene Expression Levels in Domesticated and Wild Animals," *PLOS Genetics* 8 (2012); Hare et al., in "The Self-Domestication Hypothesis" note: "An alternative evolutionary scenario to the self-domestication hypothesis is that the observed behavioural differences are due to selection for severe aggression in chimpanzees from a bonobo-like ancestor. Equally, both *Pan* species could in theory be highly derived from a common ancestor that possessed a mosaic of traits seen in both species. The ontogeny of the bonobo skull argues against these ideas. During growth, chimpanzee skulls follow closely the ontogenetic pattern of their more distant relative, gorillas, *Gorilla gorilla* . . . , whereas the bonobo cranium remains small and juvenilized compared not only to chimpanzees but also to all other great apes, including australopithecines."
16. P. Borodin, "Understanding the Person," in *Dmitry Konstantinovich Belyaev*, 2002.

17. Nikolai Belyaev, Skype interview with authors.
18. Misha Belyaev, interview with authors.
19. Misha Belyaev, interview with authors.
20. Kogan in *Dmitry Konstantinovich Belyaev*, 2002.
21. D. Belyaev, "I Believe in the Goodness of Human Nature: Final Interview with the Late D. K. Belyaev," *Voprosy Filosofii* 8 (1986): 93–94.

Chapter 8

1. A. Miklosi, *Dog Behavior, Evolution, and Cognition.*
2. By reducing activity of the adrenal cortex.
3. I. Plyusnina, I. Oskina, and L. Trut, "An Analysis of Fear and Aggression during Early Development of Behavior in Silver Foxes (*Vulpes vulpes*)," *Applied Animal Behaviour Science* 32 (1991): 253–268.
4. N. Popova, N. Voitenko, and L. Trut, "Change in Serotonin and 5-oxyindoleacetic Acid Content in Brain in Selection of Silver Foxes according to Behavior," *Doklady Akademii Nauk SSSR* 223 (1975): 1498–1500; N. Popova et al., "Genetics and Phenogenetics of Hormonal Characteristics in Animals .7. Relationships between Brain Serotonin and Hypothalamo-pituitary-adrenal Axis in Emotional Stress in Domesticated and Non-domesticated Silver Foxes," *Genetika* 16 (1980): 1865–1870.
5. More precisely, they injected foxes with L-tryptophan, a chemical precursor to serotonin.
6. A. Chiodo and M. Owyang, "A Case Study of a Currency Crisis: The Russian Default of 1998," Federal Reserve Bank of St. Louis *Review* (November/December 2002): 7–18.
7. L. Trut, "Early Canid Domestication," 168.
8. Letter from John McGrew to Lyudmila Trut.
9. Letter from Charles and Karen Townsend to Lyudmila Trut.
10. *New York Times*, February 23, 1997.
11. A nice timeline of these events can be found at the National Human Genome Research Institute website: http://unlockinglifescode.org/timeline?tid=4.

Chapter 9

1. C. Rutz and J. H. St. Clair, "The Evolutionary Origins and Ecological Context of Tool Use in New Caledonian Crows," *Behavioural Process* 89 (2013): 153–165.
2. B. Klump et al., "Context-Dependent 'Safekeeping' of Foraging Tools in New

Caledonian Crows," *Proceedings of the Royal Society B* 282 (2015), DOI:10.1098/rspb.2015.0278.

3. V. Pravosudovand and T. C. Roth, "Cognitive Ecology of Food Hoarding: The Evolution of Spatial Memory and the Hippocampus," *Annual Review of Ecology, Evolution, and Systematics* 44 (2013): 173–193.
4. J. Dally et al., "Food-Caching Western Scrub-Jays Keep Track of Who Was Watching When," *Science* 312 (2006): 1662–1665.
5. M. Wittlinger et al., "The Ant Odometer: Stepping on Stilts and Stumps," *Science* 312 (2006): 1965–1967; M. Wittlinger et al., "The Desert Ant Odometer: A Stride Integrator that Accounts for Stride Length and Walking Speed," *Journal of Experimental Biology* 210 (2007): 198–207.
6. B. Hare et al., "The Domestication of Social Cognition in Dogs," *Science* 298 (202):1634–1636. Hare did his dissertation work as a student of Richard Wrangham. His PhD dissertation was entitled "Using Comparative Studies of Primate and Canid Social Cognition to Model Our Miocene Minds" (Harvard University, 2004).
7. S. Zuckerman, *The Social Life of Monkeys and Apes* (New York: Harcourt Brace, 1932).
8. G. Schino, "Grooming and Agonistic Support: A Meta-analysis of Primate Reciprocal Altruism," *Behavioral Ecology* 18 (2007): 115–120; E. Stammbach, "Group Responses to Specially Skilled Individuals in a *Macaca fascicularis* group," *Behaviour* 107 (1988): 687–705; F. de Waal, "Food Sharing and Reciprocal Obligations among Chimpanzees," *Human Evolution* 18 (1989): 433–459.
9. A. Harcourt and F. de Waal, eds., *Coalitions and Alliances in Humans and Other Animals* (Oxford: Oxford University, 1992).
10. C. Packer, "Reciprocal Altruism in *Papio anubis*," *Nature* 265 (1977): 441–443.
11. D. Cheney and R. Seyfarth, *How Monkeys See the World* (Chicago: University of Chicago, 1990).
12. Hare's own work on this subject includes Hare et al., "The Domestication of Social Cognition"; M. Tomasello, B. Hare, and T. Fogleman, "The Ontogeny of Gaze Following in Chimpanzees, *Pan troglodytes*, and Rhesus Macaques, *Macaca mulatta*," *Animal Behaviour* 61 (2001): 335–343; S. Itakura et al., "Chimpanzee Use of Human and Conspecific Social Cues to Locate Hidden Food," *Developmental Science* 2 (1999): 448–456; M. Tomasello, B. Hare, and B. Agnetta, "Chimpanzees, *Pan troglodytes*, Follow Gaze Direction Geometrically," *Animal Behaviour* 58 (1999): 769–777; B. Hare and M. Tomasello, "Domestic Dogs (*Canis familiaris*) Use Human and Conspecific Social Cues to Locate Hidden Food," *Journal of Comparative Psychology* 113 (1999): 173–177; M. Tomasello, J. Call, and B. Hare, "Five Primate Species Follow the Visual Gaze of Conspecifics," *Animal Behaviour* 55 (1998): 1063–1069.

13. A. Miklosi et al., "Use of Experimenter-Given Cues in Dogs," *Animal Cognition* 1 (1998): 113–121; A. Miklosi et al., "Intentional Behaviour in Dog-Human Communication: An Experimental Analysis of Showing Behaviour in the Dog," *Animal Cognition* 3 (2000): 159–166; K. Soproni et al., "Dogs' (*Canis familiaris*) Responsiveness to Human Pointing Gestures," *Journal of Comparative Psychology* 116 (2002): 27–34.
14. There is an ongoing debate about wolves' ability on such tests: A. Miklosi et al., "A Simple Reason for a Big Difference"; A. Miklosi and K. Soproni, "A Comparative Analysis of Animals' Understanding of the Human Pointing Gesture," *Animal Cognition* 9 (2006): 81–93; M. Udell et al., "Wolves Outperform Dogs in Following Human Social Cues," *Animal Behaviour* 76 (2008): 1767–1773; C. Wynne, M. Udell, and K. A. Lord, "Ontogeny's Impacts on Human-Dog Communication," *Animal Behaviour* 76 (2008): E1–E4; J. Topal et al., "Differential Sensitivity to Human Communication in Dogs, Wolves, and Human Infants," *Science* 325 (2009): 1269–1272; M. Gacsi et al., "Explaining Dog/Wolf Differences in Utilizing Human Pointing Gestures: Selection for Synergistic Shifts in the Development of Some Social Skills," *PLOS ONE* 4 (2009), DOI .org/10.1371/journal.pone.0006584; B. Hare et al., "The Domestication Hypothesis for Dogs' Skills with Human Communication: A Response to Udell et al. (2008) and Wynne et al. (2008)," *Animal Behaviour* 79 (2010): E1–E6.
15. B. Hare, "The Domestication of Social Cognition in Dogs."
16. Brian Hare, Skype interview with authors.
17. B. Hare and V. Woods, *The Genius of Dogs* (New York: Plume, 2013), 78–79.
18. B. Hare et al., "Social Cognitive Evolution in Captive Foxes Is a Correlated Byproduct of Experimental Domestication," *Current Biology* 15 (2005): 226–230.
19. Other experiments were done to make sure that the foxes were not picking up olfactory cues from the hidden food.
20. Brian Hare, Skype interview with authors.
21. Forty-three tame fox pups and thirty-two control fox pups.
22. It wasn't just that control foxes were scared and uncomfortable near humans compared to tame foxes. At Brian's instruction his assistant Natalie had spent extra time with control foxes before the experiment to see to that and they ran additional experiments to be certain that was not a confounding factor.
23. Hare and Woods, 87–88.
24. Irena Muchamedshina, interview with authors.
25. R. Seyfarth, "Vervet Monkey Alarm Calls: Semantic Communication in a Free-Ranging Primate," *Animal Behaviour* 28 (1980): 1070–1094.
26. Volodin has studied communication in everything from cranes and ground squirrels to dogs and striped possums.
27. Sveta Gogoleva, email interview with authors.

28. S. Gogoleva et al., "To Bark or Not to Bark: Vocalizations by Red Foxes Selected for Tameness or Aggressiveness toward Humans," *Bioacoustics* 18 (2008): 99–132.
29. S. Gogoleva et al., "Explosive Vocal Activity for Attracting Human Attention Is Related to Domestication in Silver Fox," *Behavioural Processes* 86 (2010): 216–221.

Chapter 10

1. They also used microsatellite markers.
2. A. Kukekova et al., "A Marker Set for Construction of a Genetic Map of the Silver Fox (*Vulpes vulpes*)," *Journal of Heredity* 95 (2004): 185–194; A. Graphodatsky et al., "The Proto-oncogene C-KIT Maps to Canid B-Chromosomes," *Chromosome Research* 13 (2005): 113–122.
3. 320 loci. A. Kukekova et al., "A Meiotic Linkage Map of the Silver Fox, Aligned and Compared to the Canine Genome," *Genome Research* 17 (2007): 387–399.
4. They also compared what they had found to the genomic map of the dog. Here, what they learned was that the difference between the 17 chromosomes found in the silver fox, and the 39 typically found in dogs, was the result of various genetic fusion events. Most fox chromosomes were made up of chunks of more than one dog chromosome.
5. National Institute of Mental Health, Molecular Mechanisms of Social Behavior, MH0077811, 08/01/07–07/31/11; National Institute of Mental Health, Molecular Genetics of Tame Behavior MH069688, 04/01/04–03/31/07.
6. K. Chase et al., "Genetic Basis for Systems of Skeletal Quantitative Traits: Principal Component Analysis of the Canid Skeleton," *Proceedings of the National Academy of Sciences of the United States of America* 99 (2002): 9930–9935; D. Carrier, K. Chase, and K. Lark, "Genetics of Canid Skeletal Variation: Size and Shape of the Pelvis," *Genome Research* 15 (2005): 1825–1830.
7. K. Chase et al., "Genetic Basis for Systems of Skeletal Quantitative Traits"; L. Trut et al., "Morphology and Behavior: Are They Coupled at the Genome Level?" in *The Dog and Its Genome*, ed. E. A. Ostrander, U. Giger, and K. Lindblad-Toh (Woodbury, NY: Cold Spring Harbor Laboratory Press, 2005), 81–93.
8. Using mathematical models developed by geneticists, Anna and Lyudmila constructed a very specific breeding protocol that involved mating tame and aggressive foxes with each other over the course of three generations, so that the molecular genetic analysis would have the maximum power to locate any genes associated with tame behavior; A. Kukekova et al., "Measurement of

Segregating Behaviors in Experimental Silver Fox Pedigrees," *Behavior Genetics* 38 (2008): 185–194.

9. A. Kukekova et al., "Sequence Comparison of Prefrontal Cortical Brain Transcriptome from a Tame and an Aggressive Silver Fox (*Vulpes vulpes*)," *BMC Genomics* 12 (2011): 482, DOI:10.1186/1471-2164-12-482. Preliminary work done here includes J. Lindberg et al., "Selection for Tameness Modulates the Expression of Heme Related Genes in Silver Foxes," *Behavioral and Brain Functions* 3 (2007), DOI:10.1186/1744-9081-3-18; J. Lindberg et al., "Selection for Tameness Has Changed Brain Gene Expression in Silver Foxes," *Current Biology* 15 (2005): R915–R916.
10. Back in his day, Belyaev had suggested something else about gene expression and domestication. He proposed that large clusters of genes whose expression affects the process of domestication might themselves be under the control of a select few "master regulatory genes." If correct, these master regulatory genes might then control many of the changes that have emerged during fox domestication—changes in behavior, coat color, hormone level, bone length and width, and so on—all at once. Lyudmila and Anna know that finding these master regulatory genes, should they even exist, is years off in the future. But, when it comes to her beloved foxes, Lyudmila has become something of an expert on planning for things that seem far, far off in the future. If they could eventually find these master regulatory genes that control gene expression in clusters of other genes, and sequence them, Lyudmila believes that the fox team might just tap into "control [of] the entire domestication process."
11. The genes were *SOX6* and *PROM1*: F. Albert et al., "A Comparison of Brain Gene Expression Levels in Domesticated and Wild Animals," *PLOS Genetics* 8 (2012), doi.org/10.1371/journal.pgen.1002962.
12. M. Carneiro et al., "Rabbit Genome Analysis Reveals a Polygenic Basis for Phenotypic Change during Domestication," *Science* 345 (2014): 1074–1079.
13. A. Wilkins, R. Wrangham, and T. Fitch, "The 'Domestication Syndrome' in Mammals: A Unified Explanation Based on Neural Crest Cell Behavior and Genetics," *Genetics* 197 (2014): 795–808.
14. Letter from Rene and Mitchell to Lyudmila Trut.
15. Letter from Moiseev Dmitry to Lyudmila Trut.

Index